# Image and Video Processing in the Compressed Domain

# Image and Video Processing in the Compressed Domain

Jayanta Mukhopadhyay

CRC Press
Taylor & Francis Group
Boca Raton   London   New York

CRC Press is an imprint of the
Taylor & Francis Group, an **informa** business

A CHAPMAN & HALL BOOK

Chapman & Hall/CRC
Taylor & Francis Group
6000 Broken Sound Parkway NW, Suite 300
Boca Raton, FL 33487-2742

First issued in paperback 2017

© 2011 by Taylor and Francis Group, LLC
Chapman & Hall/CRC is an imprint of Taylor & Francis Group, an Informa business

No claim to original U.S. Government works

ISBN 13: 978-1-138-11378-7 (pbk)
ISBN 13: 978-1-4398-2935-6 (hbk)

Dedicated to my parents.

# Preface

With more and more images and videos being available in the compressed format, researchers have started taking interest in the aspect of designing algorithms for different image operations directly in their domains of representation. This would not only avoid the inverse and forward transformation steps and potentially make the computation faster but also would keep the buffer requirement less as the storage required for processing the compressed stream is less than that with its uncompressed representation.

This book attempts to comprehensively treat this topic of interest and deals with the fundamentals and properties of various image transforms used in image and video compression. Subsequently, their application in designing image and video processing algorithms in the compressed domain are discussed. To provide better understanding of the domain of research, different image and video compression techniques are briefly covered in the first chapter. In particular, discrete cosine transform (DCT) - based compression algorithms (such as JPEG, MPEG, and H.264) and discrete wavelet transform (DWT) - based JPEG2000 are discussed with more details. This is followed by discussion on key properties of various transform spaces with special emphasis on the block DCT space and the DWT.

Different types of image and video processing operations performed in the compressed domain are discussed in subsequent chapters. This includes filtering, enhancement and color restoration, image and video resizing, transcoding, etc. In addition, different other applications in the compressed domain such as video and image editing, digital watermarking and steganography, image and video indexing, face detection and identification, etc., are briefly covered in the last chapter.

This book is meant for readers who have gone through a first-level course on digital image processing and are familiar with the basic concepts and tools of image and video processing. However, at the introductory level, the details of compression techniques and properties of the transform domain are not always extensively covered. The first two chapters of this book discuss these issues at considerable depth. For the sake of completeness, an effort has also been made to develop concepts related to compressed domain processing from the very basic level so that a first-time reader also does not have any problem in understanding them.

The book has seven chapters. In the first chapter, the motivation and background for processing images and videos in the compressed domain are discussed. There is also a brief introduction to different popular image and video compression algorithms, notably JPEG, JPEG2000, MPEG-2, MPEG-4, and H.264 standards for lossy image and video compression schemes. Issues related to compressed domain analysis and performance metrics for comparing different algorithms are also elaborated in this chapter.

The second chapter elucidates the definitions and properties of different image transforms, in particular, the discrete fourier transform (DFT), DCT, integer cosine transform (ICT), and DWT. The last three transforms are chosen because of their use in compression technologies. In subsequent chapters, some of the core operations such as filtering, resizing, etc., which find use in various approaches in image and video analysis exploiting these properties, are discussed.

The third chapter considers image filtering in the block DCT domain. In this chapter the convolution and multiplication properties of DCTs are elaborated, followed by discussion on different approaches for computing the filtered response directly in the compressed domain. Typical applications of filtering are also briefly presented in this chapter.

In chapter four, with a general introduction to color processing in the compressed domain, a few representative problems are considered. The first one is related to image enhancement and restoration through saturation and desaturation of colors. In the second case, the problem of color constancy is introduced, and various approaches for solving this problem in spatial and compressed domain are illustrated. A comparative study of different representative schemes is also presented here. Next, the problem of enhancing colors in the block DCT domain is taken into account, and different algorithms for performing this task are discussed.

Chapter five focusses on the image resizing problem in the block DCT space. Different approaches are discussed in this regard. Initially, various techniques for image halving and doubling are discussed. Later, the problem of arbitrary resizing is considered. The chapter also introduces the concept of hybrid resizing and discusses its solution in the block DCT space. The problem of video resizing, in particular video downsampling, is discussed in the next chapter on transcoding.

In chapter six, transcoding of images and videos is discussed. As most of the image and video standards use the DCT, ICT, or DWT for their representation, first, techniques for intertransform conversion are discussed. These are followed by a discussion on various types of transcoding operations. The topics of interest include transcoding of a JPEG2000 image into JPEG, an H.264 video into MPEG-2, and vice versa. The discussion is facilitated with the introduction of different measures related to the performance of a transcoder. The chapter also discusses techniques for altering temporal and spatial resolution of videos by skipping frames at regular intervals and reducing their frame sizes, respectively. At the end of the chapter, error-resilient transcoding of the video stream is also discussed.

There are various other applications of processing of images and videos in the compressed domain. They include different video editing operations such as key frame extraction, caption localization, object recognition, etc. There are also different methods of indexing videos using features computed from the block DCT and DWT spaces. Image and video steganography and watermarking in the compressed domain are also major topics of research in the area of multimedia security. All these different facets of compressed domain analysis are put together in the concluding seventh chapter.

Even after going through several revisions of the text, I always found scopes for improvements at every iteration. Finally, I had to settle for this version to meet the deadline and other commitments. I would greatly appreciate if the readers of this book, after encountering errors in the printed text, would bring them to my notice.

While working in this area I have been fortunate to have had the guidance and friendship of Professor Sanjit K. Mitra of the University of Southern California, Los Angeles. I take this opportunity to express my deepest gratitude and respect for his constant encouragement and enlightenment. I am also thankful to my colleagues Professor P.K. Biswas and Professor Rajeev Kumar of IIT, Kharagpur, who have worked with me in this area of research. My gratitude also goes to my former students Dr. K. Viswanath, Dr. V. Patil, Mr. Sudhir Porwal, and Ms. T. Kalyani, who contributed in this area at different stages and enriched my understanding of this topic. Professor Shamik Sural of IIT, Kharagpur, went through several versions of this book and greatly helped in improvising the present edition. Without his help and constant encouragement, it would not have been possible for me to complete this book. I also thank Dr. Sreepat Jain of CRC press who invited me to write this book and initiated the project. Later, I received able support from Ms. Aastha Sharma and Ms. Jessica Vakili toward its completion. I am grateful to Jim McGovern who did an extensive proof reading of the manuscript and helped me to correct many typos and grammars in this book. Though I am pretty sure my wife Jhuma and my son Rudrabha will be least interested in reading the book's content, they would be at least happy to see the end of their nightmare due to my late-night intellectual exercises to meet the submission-deadline. I especially thank them for their support, patience, and understanding. Finally, I dedicate this book to my parents, from whom I learned my first lessons with the greatest joy ever I had in my life.

Jayanta Mukhopadhyay
24th September, 2010
IIT Kharagpur, India

# List of Figures

xi

# List of Tables

# Contents

# Symbol Description

Notations are defined here in 1-D. The same notations are also used in their extended definitions in 2-D.

## Sets and spaces

| | |
|---|---|
| $\mathbb{R}$ | The set of all real numbers. |
| $\mathbb{Z}$ | The set of all integers. |
| $\mathbb{N}$ | The set of all non-negative integers. |
| $\mathbb{Z}_N$ | $\{0, 1, 2, \ldots, N-1\}$ |
| $\mathbb{C}$ | The set of all complex numbers. |
| $L^2(\mathbb{R})$ | The space of all square integrable functions. |
| $L^2(\mathbb{Z})$ | The space of all square integrable functions over the integer grid. |
| $[a, b]$ | $\{x \mid x \in \mathbb{R} \text{ and } a \leq x \leq b\}$ |

## Functions, sequences, operators and symbols

| | |
|---|---|
| $<h, g>$ | The inner product of two functions $h(x)$ and $g(x)$. |
| $\vec{h} \cdot \vec{g}$ | The dot product of two vectors $\vec{h}$ and $\vec{g}$. |
| $a + jb$ | A complex number with $j$ as $\sqrt{-1}$. |
| $x^*$ | The complex conjugate of $x \in \mathbb{C}$. |
| $\|x\|$ | The magnitude of $x \in \mathbb{C}$. |
| $\angle x$ | The phase of $x \in \mathbb{C}$. |
| $<x>_N$ | $x \bmod N$ for $x \in \mathbb{Z}$. |
| $\|x\|$ | The absolute value of $x \in \mathbb{R}$. |
| $sign(x)$ | $-1$, $0$, and $1$ depending on the sign of $x \in \mathbb{R}$. |
| $round(x)$ | The nearest integer approximation of $x \in \mathbb{R}$. |
| $\lfloor x \rfloor$ | The nearest integer, which is less than or equal to $x \in \mathbb{R}$. |
| $\lceil x \rceil$ | The nearest integer, which is greater than or equal to $x \in \mathbb{R}$. |
| $f(x)$ | A continuous function ($x \in \mathbb{R}$) in $L^2(\mathbb{R})$. |
| $\delta(x)$ | The Dirac delta function. |
| $f(n)$ | A discrete function ($n \in \mathbb{Z}$) in $L^2(\mathbb{Z})$. |
| $f^+(n)$ | Positive half of $f(n)$ for $n \geq 0$. |
| $f^P(n)$ | Strict positive half of $f(n)$ for $n > 0$. |
| $x_{de}(m)$ | Even down-sampled sequence of $x(n)$. |
| $x_{do}(m)$ | Odd down-sampled sequence of $x(n)$. |
| $x_{ue}(m)$ | Even up-sampled sequence of $x(n)$. |
| $x_{uo}(m)$ | Odd up-sampled sequence of $x(n)$. |
| $\overline{w(n)}$ | The conjugate reflection of $w(n) \in \mathbb{C}$. |
| $\|\mathbf{x}\|$ | The Euclidean norm of the vector $\mathbf{x}$. |
| $f \star h(n)$ | Linear convolution of $f(n)$ and $h(n)$ ( or $f(n) \star h(n)$). |
| $f \circledast h(n)$ | Circular convolution of $f(n)$ and $h(n)$ ( or $f(n) \circledast h(n)$). |
| $f \circledS h(n)$ | Skew circular convolution of $f(n)$ and $h(n)$ ( or $f(n) \circledS h(n)$). |
| $f \boxplus h(n)$ | Symmetric convolution of $f(n)$ and $h(n)$ ( or $f(n) \boxplus h(n)$). |
| $n_m$ | Number of multiplications. |
| $n_a$ | Number of additions. |

## Transforms

| | |
|---|---|
| $\mathbb{F}(f(x))$ | Fourier transform of $f(x)$. |
| $\hat{f}(j\omega)$ | Fourier transform of $f(x)$ or $\mathbb{F}(f(x))$. |
| $\|\hat{f}(j\omega)\|$ | Magnitude spectrum of $f(x)$. |
| $\theta(\omega)$ | Phase spectrum of $f(x)$. |
| $\mathbb{F}(f(n))$ | The DFT of $f(n)$. |
| $\hat{f}(k)$ | The DFT of $f(n)$ or $\mathbb{F}(f(n))$. |
| $\mathbb{F}_{\alpha,\beta}(f(n))$ | The GDFT of $f(n)$ for $\alpha, \beta \in \{0, \frac{1}{2}\}$. |
| $\hat{f}_{\alpha,\beta}(k)$ | The GDFT of $f(n)$ for $\alpha, \beta \in \{0, \frac{1}{2}\}$, or $\mathbb{F}_{\alpha,\beta}(f(n))$. |
| $\hat{\vec{x}}$ | The DFT of $\vec{x}$. |
| $\hat{f}_{0,\frac{1}{2}}(k)$ | The Odd Time Discrete Fourier Transform ($OTDFT$) of $f(x)$. |
| $\hat{f}_{\frac{1}{2},0}(k)$ | The Odd Frequency Discrete Fourier Transform ($OFDFT$) of $f(x)$. |
| $\hat{f}_{\frac{1}{2},\frac{1}{2}}(k)$ | The Odd Frequency Odd Time Discrete Fourier Transform ($O^2DFT$) of $f(x)$. |
| $H(z)$ | The z-transform of $h(n)$. |
| $C_{ie}(x(n))$ | Type-i even DCT of $x(n)$ for $i \in \{1, 2, 3, 4\}$. |
| $X_{ie}(k)$ | Type-i even DCT of $x(n)$ for $i \in \{I, II, III, IV\}$ (an alternative notation of $C_{ie}(x(n))$). |
| $C_{io}(x(n))$ | Type-i odd DCT of $x(n)$ for $i \in \{1, 2, 3, 4\}$. |
| $X_{io}(k)$ | Type-i odd DCT of $x(n)$ for $i \in \{I, II, III, IV\}$ (an alternative notation of $C_{io}(x(n))$). |
| $S_{ie}(x(n))$ | Type-i even DST of $x(n)$ for $i \in \{1, 2, 3, 4\}$. |
| $X_{ise}(k)$ | Type-i even DST of $x(n)$ for $i \in \{I, II, III, IV\}$ (an alternative notation of $S_{ie}(x(n))$). |
| $S_{io}(x(n))$ | Type-i odd DST of $x(n)$ for $i \in \{1, 2, 3, 4\}$. |
| $X_{iso}(k)$ | Type-i odd DST of $x(n)$ for $i \in \{I, II, III, IV\}$ (an alternative notation of $S_{io}(x(n))$). |
| $DCT(x)$ | Type II even DCT of $x$. |
| $DST(x)$ | Type II even DST of $x$. |

## Matrices and operators

| | |
|---|---|
| $X^T$ | The transpose of matrix $X$. |
| $X^H$ | The Hermitian transpose of matrix $X$. |
| $X^{-1}$ | The inverse of matrix $X$. |
| $A \otimes B$ | Element wise multiplication of $A$ and $B$. |
| $[f(k, l)]$ | The matrix formed in such a way that its $(k, l)$th element is $f(k, l)$. |

$\mathbf{x}$ — The column vector formed from $x(n)$ such that $i$th element of the vector is $x(i)$.

$\{\mathbf{x}\}_p^q$ — The column vector formed from $\mathbf{x}$ from its $p$th element to $q$th one.

$\mathbb{D}(\mathbf{x})$ — The diagonal matrix whose $(i, i)$th diagonal element is $x(i)$.

$\mathbb{D}_m(\mathbf{x})$ — The diagonal matrix whose $m$th off diagonal elements are formed from $\mathbf{x}$ in the order of appearances while scanning from left to right and top to bottom.

$\Phi_N$ — The $N \times N$ flipping matrix.

$\Psi_N$ — The diagonal matrix $\mathbb{D}(\{(-1)^m\}_{m=0}^{N-1})$.

$\mathbf{F}$ — The DFT matrix.

$\mathbf{F}_{\alpha,\beta}$ — The GDFT matrix for $\alpha, \beta \in \{0, \frac{1}{2}\}$.

$C_N^\alpha$ — $N$-point Type $\alpha$ even DCT matrix, where $\alpha \in \{I, II, III, IV\}$.

$S_N^\alpha$ — $N$-point Type $\alpha$ even DST matrix, where $\alpha \in \{I, II, III, IV\}$.

$C_N$ — $N$-point Type II even DCT matrix.

$S_N$ — $N$-point Type II even DST matrix.

$C_8$ — 8-point Type II even DCT matrix.

$T_8$ — 8-point ICT matrix.

$T_4$ — 4-point ICT matrix.

$Hd_m$ — Hadamard matrix of size $2^m \times 2^m$.

$\mathbb{H}_N$ — Discrete Haar Transform matrix of size $N \times N$.

$P_N$ — $N \times N$ permutation matrix.

$0_N$ — $N \times N$ zero or null matrix.

$0_{M \times N}$ — $M \times N$ zero or null matrix.

$I_N$ — $N \times N$ identity matrix.

$J_N$ — $N \times N$ reverse identity matrix.

$A_{M,N}$ — DCT block composition matrix for merging $M$ adjacent $N$-point DCT blocks.

$B_{M,N}$ — DST block composition matrix for merging $M$ adjacent $N$-point DCT blocks into a DST block.

# Chapter 1

# Image and Video Compression: An Overview

1

Presently we are witnessing an explosion of information. Information is generated in various forms such as text, audio, image, video, etc., which are commonly referred to as multimedia content. Within a very short period, availability of inexpensive yet powerful digital cameras and video recorders in the consumer world has further accelerated the pace of research and development in processing these media. One of the very initial tasks of processing these data is to compress them prior to their storage and/or transmission. By compressing data, we want to reduce the size required for their representation. For image and video, this is usually achieved by an alternative representation in a different domain other than their original domains of representation, which are spatial and spatiotemporal for images and videos, respectively. Two such popular representations of images and videos are formed by their *discrete cosine transforms* (DCT) [3] and *discrete wavelet transforms* (DWT) [92]. We refer to these domains as *compressed domains* or *domains of compression*. Since images and videos are increasingly available in these forms, there is a need for their direct analysis in those spaces. This potentially makes the processing faster by avoiding the overhead of decompression and recompression of the data. Further, a smaller data size in the compressed domain provides an additional advantage of keeping the memory requirement low. However, we should make a careful cost-benefit analysis in terms of actual computation cost and memory requirement before suggesting any equivalent processing in these domains. Even though direct processing in compressed domain avoids the decompression and recompression tasks, equivalent computations with transform coefficients may become too costly compared to their spatial domain operations.

## 1.1   Compression: Generic Approaches

Images are represented by a function over a 2-D integral coordinate space such as $I : \mathbb{Z}^2 \rightarrow \mathbb{R}$, where $\mathbb{Z}$ and $\mathbb{R}$ are a set of integers and real numbers, respectively. Each sample in the 2-D integral coordinate space is referred to as a *pixel*. Usually, functional values at pixels are quantized to a nonnegative integer (let the set be denoted by $\mathbb{N}$), implying proportional brightness distribution over the 2-D space. For a color image, every pixel is mapped to a three-dimensional color vector in the RGB color space ($f : \mathbb{Z}^2 \to \mathbb{N}^3$). A video, on the other hand, is represented by a temporal sequence of images that are observed at evenly spaced discrete time points, and its function can be represented as $V : \mathbb{Z}^2 \times \mathbb{N} \to \mathbb{N}$, and for a colored video $V_c$, the mapping is expressed as $V_c : \mathbb{Z}^2 \times \mathbb{N} \to \mathbb{N}^3$.

### 1.1.1 Alternative Representation

While compressing images or videos, we look for an alternative representation that requires less storage space compared to their representations in the original space. Here we may draw an analogy with the representation of a circle. A circle may be represented by a set of points in the 2-D coordinate space. However, only three points are sufficient to uniquely define (or represent) the circle. So, given $m$ ($m \geq 3$) points on a circle, ($m - 3$) points are redundant, and we may reduce the storage requirement by removing those redundancies from the representation. Alternatively, we may represent a circle by its center and radius, which also require less storage than its representation as a set of points in the 2-D space. Moreover, the representation may not always be exact. Consider the representation of closed curves or contours by approximate circles or ellipses. In such cases, there is a significant amount of savings in the storage requirement. In some applications, this type of approximation may be acceptable. In fact, for images and videos, many a time we work with approximate representations of original images captured by cameras. Compression schemes dealing with approximate representations of objects are known as **lossy compression schemes**, while the schemes with exact representation are known as **lossless schemes**.

For any compression scheme, its alternative representation of images (or data) should have the following features.

1. **Reconstructibility**: Alternative representation is a form of encoding of data. We should be able to decode or reconstruct the data from its encoded form. This reverse process is known as reconstruction. However, the reconstruction may be partial or approximate for a lossy compression.

2. **Low redundancy**: The representation should have low redundancy in its information content. This redundancy may occur in various forms. For images there are redundancies due to spatial correlation among pixels. Their color components are also correlated, in particular, in the RGB color space. For videos, in addition, there are temporal correlations among their consecutive frames. Besides spatial and temporal correlations, other kinds of redundancies are also observed. Sometimes these

are context dependent, such as symmetry of objects, biased distribution of brightness and colors, etc.

3. **Factorization into substructures**: Decomposition of the object into its different components or substructures is another desirable feature of a representation scheme. This becomes particularly useful for approximate or lossy representation of objects. In this case, components contributing insignificantly to the reconstruction of an object in its original form, are pruned or removed. This reduces the storage requirement of the representation.

In various ways, these alternative representations are possible. We may use image transforms to represent images by a set of coefficients corresponding to a given set of basis functions. These coefficients may be real numbers (e.g., DCT, DWT, etc.) or complex numbers (*discrete Fourier transform* (DFT) [100]). Another form of image representation is the fractal representation with *iterated function systems* (IFS), in particular *partitioned iterated function systems* (PIFS) [65] . In this technique, images are represented by a set of *contractive affine transformations*, which, being applied iteratively to an initial configuration, converge to the desired image. Color components may also be transformed into other color spaces from the usual RGB color space so that they become less correlated, requiring less number of bits for representation. In some cases, one color component may be expressed as functions of others, requiring less storage in representing those functions.

Though each frame of a video may be represented independently using any of the image representation techniques, to exploit the temporal coherence existing among consecutive frames, we may consider representing temporal offsets between them. In this technique, considering one of the frames as a reference frame among a *group of pictures* (GOP), other frames are described by temporal offsets (or motions) of their different parts. Usually, the first frame of the GOP is taken as a reference frame and called *Intra Frame* or *I-Frame*. As other frames are predicted from it, they are called *P-Frames*. Some of them may also be predicted from two neighboring frames using bidirectional prediction in the temporal domain. They are referred to as *B-Frames*.

In some representations, this temporal cohesion is not explicitly represented (in the form of motion vectors and prediction errors). In this case, some of the frames are represented as side information, and others are encoded by a set of parity bits in such a way that, given side information, we would be able to approximately decode the frames from them. This kind of representation is specially used for a low-complexity encoder and in a distributed environment, where each frame is encoded independently. This type of encoding is referred to as *distributed video coding* [51].

## 1.1.2 Quantization

Alternative representation of images and videos implies that they are represented by a different set of symbols, messages, or information units. For example, image transforms are used to represent images by a set of coefficients. In a video, temporal sequences may be represented by differences of pixel values at the same location of consecutive frames. This derivation of representation is the first stage of compression. Sometimes symbols used in describing this representation, are termed *intermediate symbols*. In lossy compression techniques, often values of intermediate symbols are scaled down by quantization parameters so that, the dynamic ranges of those symbols get reduced and thus they require less number of bits for their representations. These quantization parameters may vary among the components of the alternative representation. For images, videos, and audio, the values of these parameters are recommended by observing the effect of individual components on average human perception.

## 1.1.3 Entropy Coding

At the final stage of compression, we should target the efficient encoding of intermediate symbols considering their statistical distributions in the domain of application. As the techniques developed for these tasks are guided by *information theory* [52], they are commonly referred to as *entropy encoding*. In Figure 1.1, two important steps of compression and decompression are shown.

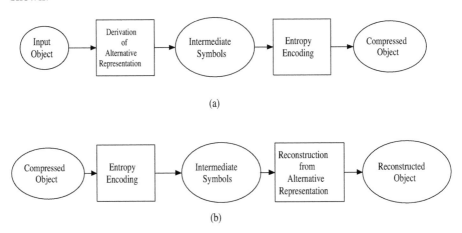

(a)

(b)

**Figure 1.1**: Schematic representations of (a) compression and (b) decompression schemes.

Image transforms such as DCT and DWT, are widely used for representing images with transform coefficients. They have the desirable properties of reconstructibility, low redundancy, and factorization in representation of images. Due to the last property, they are quite useful in approximating images with less number of coefficients, which inherently reduces the storage requirement. Further, by employing entropy coding techniques such as Huffman encoding [52], arithmetic encoding [52], etc., intermediate symbols are represented with less number of bits.

### 1.1.4   Rate-Distortion Control

In lossy compression, the higher the values of quantization parameters, the less is the required number of bits of representation. However, high quantization introduces more error in the representation of intermediate symbols, leading to more distortion in the reconstructed image or video. We should judge the performance of a compressor by these two figures of merit, that is, *rate* or normalized size of the data, and *distortion* or the average error of reconstruction. The higher the rates, the lower the distortion is. However, the rate should not be greater than that of lossless representation of the object (nor should it go below zero!). By varying different parameters such as quantization thresholds for different factors, we could achieve varying rate distortion performance. The objective of rate-distortion control is to achieve minimum distortion at a given rate. Alternatively, the problem can also be formulated as minimization of rate given a distortion.

---

## 1.2   Motivation for Processing in the Compressed Domain

For processing compressed images and videos, we first need to decompress them and apply suitable processing algorithms, which are usually developed considering their representations in spatial or spatiotemporal domains. Often, it is also necessary to recompress the processed output, to make them interoperable with other applications. For example, a JPEG compressed image, after processing, may still require to be represented by the JPEG compression standard [149] for its display by a browser, storage in an archive, or transmission through a network. Under this scenario, every spatial domain processing of images is encapsulated by additional overheads of decompression and compression as depicted in Figure 1.2.

To reduce the overhead involved in the decompression and recompression steps (Figure 1.2), we may consider performing processing with the intermediate symbols of alternative representation of images in the compressed form instead of working with their original representations in the spatial domain. For example, in a DCT-based image representation, we may compute

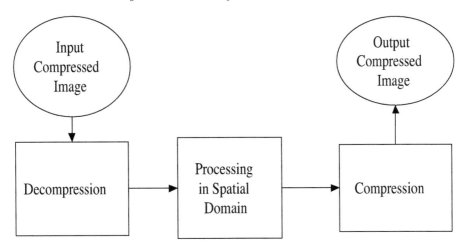

Figure 1.2: Basic steps in processing with compressed images.

with the DCT coefficients of the image and produce the DCT coefficients
of the processed images. In this case, though there are overheads of entropy
decoding and encoding of intermediate symbols (e.g., DCT coefficients, see
Figure 1.3), the total overhead is significantly less than that of decompression
and recompression. *This processing in the domain of alternative representation is referred here as processing in the compressed domain.* However, we
should also be concerned about the efficiency of the processing algorithm in
the compressed domain. If this algorithm takes longer time than its equivalent
operation in the spatial domain, the saving in computation of reconstruction
and derivation of alternative representation (or *inverse transform* and *forward
transform* operations of transform coding) are not cost effective.

There is also another advantage of working with the representation in the
compressed domain. Usually, the size of data in this case is smaller than that
of its spatial representation (or spatiotemporal representation of videos). This
inherently makes the processing less memory intensive.

Third, the property of factorization of an image in its alternative representation sometimes becomes useful in developing algorithms. We may consider
emphasizing or deemphasizing some factors after analyzing their role in the
reconstruction of the image.

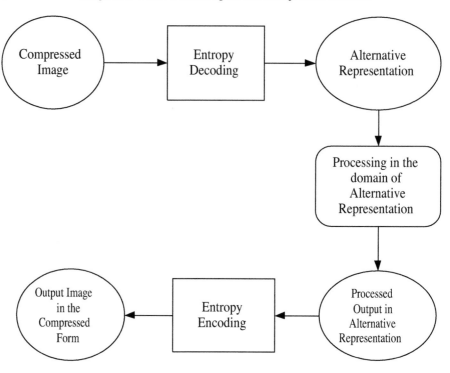

Figure 1.3: Processing in the domain of alternative representation.

## 1.3 Overview of Different Image and Video Compression Techniques and Standards

Various image and video compression techniques have been advanced by researchers starting from the early eighties. Different international bodies such as the International Standard Organization (ISO) and the International Telecommunication Union (ITU) have been coordinating the standardization task under their various committees. One such committee, Joint Photographic Experts Group (JPEG), was created by the ISO in 1986, and it recommended a still image compression standard based on DCT in 1992, later approved in 1994 as ISO 10918-1. The standards adopted for lossy and lossless image compression are commonly referred to as the JPEG [149] and JPEG-LS compression standards, respectively. Later in the year 2000, the same committee recommended another image compression standard based on DWT for both lossy and lossless image compression. This latter compression standard is named JPEG2000 [27].

A similar initiative has also been taken for standardizing video compression methods. Initially, both the ISO and the ITU took on this task independently. The ISO formed a committee named the Motion Picture Experts Group (MPEG) to set audio and video standards around 1988. Since then the committee has been recommending different video compression standards for different ranges of data rates and video quality such as MPEG1, MPEG2, and MPEG4. The MPEG-1 [47] standard is targeted at supporting data rates of a compact disk (CD), whereas MPEG-2 [135] is targeted at applications with higher bit rates. The MPEG-4 [135] standard added more features in the process of compression for achieving a higher compression rate and making the standard suitable for both high and low bit rate applications.

The ITU (then the CCITT) initiated the standardization of video compression prior to the ISO. They have recommended different H.26x video coding standards, out of which the first one, H.261, was published in 1990, and it greatly influenced the subsequent video coding standards. Later, this standard evolved as H.263 and H.264 in 1998 and 2003, respectively. The committee looking after the development of these compression standards is known as the Video Coding Expert Group (VCEG) (formerly called the ITU-T). In 2003 they formed a Joint Video Team (JVT) with MPEG, and the JVT is presently carrying out the extension of H.264 incorporating various other features.

As images and videos are mostly available in one of the above-mentioned compression standards, we will review a few of them here. In particular, we will pay attention to lossy JPEG and both lossy and lossless JPEG2000 compression standards for still images. For videos, we will discuss MPEG-2, MPEG-4, and H.264 standards [155].

## 1.4  Image Compression Techniques

In the JPEG image compression standard, there are different modes of operations,

1. Lossy sequential encoding scheme

2. Lossy progressive encoding scheme

3. Lossy hierarchical encoding scheme

4. Lossless (sequential) encoding scheme.

Out of this, first three are based on DCT-based representation of images. The last one, the lossless JPEG compression scheme (referred to as JPEG-LS in the literature) is in the form of differential encoding after prediction from neighboring pixels. Usually, JPEG compression of images refers to lossy sequential encoding (sometimes called the "baseline JPEG compression algorithm"). The progressive encoding scheme considers encoding of the details of an image in multiple scan so that images are progressively reconstructed during a long transmission session. Hierarchical encoding enables ready access to lower resolution representation of images without going for reconstruction with their full (or higher) resolutions. We restrict our discussion to the baseline sequential encoding algorithm, as this is the basis of image representation for other modes of operations of lossy JPEG compression.

### 1.4.1  Baseline Sequential JPEG Lossy Encoding Scheme

In this scheme, an image is partitioned into $8 \times 8$ blocks, and each block is almost independently coded. In Figure 1.4 the different stages of compression are shown. In this compression scheme, an image is scanned from left to right and top to bottom. In this order, each $8 \times 8$ nonoverlapping block is subjected to a set of operations. These are briefly discussed in following subsections.

#### 1.4.1.1  Level Shifting

In the JPEG compression standard, each pixel of a monochrome image has an 8-bit grey-level value (for color images each pixel is of 24 bits, that is, 8 bits for each red (R), green (G), and blue (B) component). To keep the dynamic range between $-128$ to 127, 128 is subtracted from each pixel.

#### 1.4.1.2  Transformation

The level-shifted values in each block are transformed by applying an $8 \times 8$ DCT. Let the pixels in a block be denoted by $x(m,n), 0 \leq m, n \leq 7$. The DCT of these pixels $X(k,l), 0 \leq k, l \leq 7$ are computed as follows.

$$X(k,l) = \tfrac{1}{4}.\alpha(k).\alpha(l). \sum_{m=0}^{7} \sum_{n=0}^{7} (x(m,n)cos(\frac{(2m+1)\pi k}{16})cos(\frac{(2n+1)\pi l}{16})),$$

$$0 \leq k, l \leq 7, \tag{1.1}$$

where $\alpha(k)$ is given by

$$\alpha(k) = \sqrt{\tfrac{1}{2}}, \quad \text{for } k = 0$$
$$= 1, \quad \text{otherwise.} \tag{1.2}$$

From $X(k,l), 0 \leq k, l \leq 7$, $x(m,n)$ is completely recovered by applying the inverse transform as follows.

$$x(m,n) = \tfrac{1}{4}. \sum_{k=0}^{7} \sum_{l=0}^{7} (\alpha(k).\alpha(l).X(k,l)cos(\frac{(2k+1)\pi m}{16})cos(\frac{(2l+1)\pi n}{16})),$$

$$0 \leq m, n \leq 7. \tag{1.3}$$

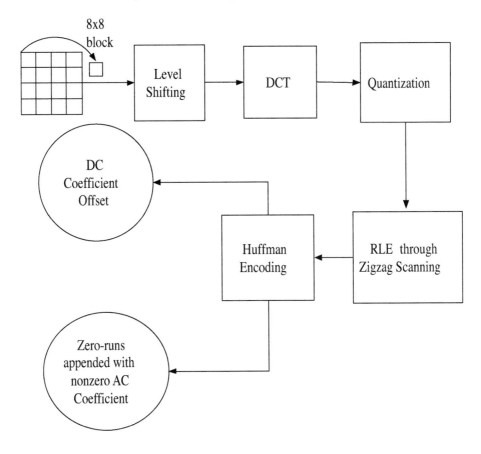

Figure 1.4: Steps in baseline sequential JPEG lossy compression scheme.

### 1.4.1.3 Quantization

In the next step, DCT coefficients of the block are quantized using different quantization thresholds for different frequency components. Let $X(k,l)$ be the $(k,l)$th DCT coefficient (for the spatial frequencies at $(k,l)$), and let $Q(k,l)$ be the corresponding quantization threshold. Then, the quantized transform coefficient is computed as

$$X_q(k,l) = round\left(\frac{X(k,l)}{Q(k,l)}\right), \qquad (1.4)$$

where $round(x)$ is the nearest integer approximation of $x$.

During decoding, approximate DCT coefficients are obtained by multiplying the corresponding quantization threshold with the quantized coefficient as follows:

$$\hat{X}(k,l) = X_q(k,l).Q(k,l). \qquad (1.5)$$

The process of quantization–dequantization introduces loss in the reconstructed image and is inherently responsible for the "lossy" nature of the compression scheme. However, quantization reduces the dynamic range of values of a transform coefficient, thus requiring less number of bits for its representation. Moreover, many insignificant values are masked by quantization thresholds and are set to zero. This happens more often for high-frequency coefficients. These runs of zeroes are efficiently coded in the later stage of compression.

Conducting various psychovisual experiments on different sets of images, a few quantization tables have been recommended by JPEG. However, we can also define a new quantization table and use it by providing the necessary information in the header of the compressed stream or file. A typical quantization table is shown in Figure 1.5.

| 16 | 11 | 10 | 16 | 24 | 40 | 51 | 61 |
|----|----|----|----|----|-----|-----|-----|
| 12 | 12 | 14 | 19 | 26 | 58 | 60 | 55 |
| 14 | 13 | 16 | 24 | 40 | 57 | 69 | 56 |
| 14 | 17 | 22 | 29 | 51 | 87 | 80 | 62 |
| 18 | 22 | 37 | 56 | 68 | 109 | 103 | 77 |
| 24 | 35 | 55 | 64 | 81 | 104 | 113 | 92 |
| 49 | 64 | 78 | 87 | 103 | 121 | 120 | 101 |
| 72 | 92 | 95 | 98 | 112 | 100 | 103 | 99 |

Figure 1.5: A typical quantization table.

Using a single quantization table, we may proportionally vary all the quantization thresholds to control the degree of quantization. In this case, quantization thresholds are scaled by a constant factor.

### 1.4.1.4   Encoding DC Coefficients

The $(0,0)$th frequency component of the block (i.e., $X(0,0)$ in Eq.(1.1)) is called the *DC coefficient*, as it is equivalent to the average intensity value of the block. The rest of the coefficients for all other frequency components are known as *AC coefficients*. Usually, DC coefficients of neighboring blocks are highly correlated. That is why these coefficients are separately encoded by computing offsets between adjacent blocks (see Figure 1.6). Given the initial value in a sequence of offsets, the true values of all subsequent coefficients are derived.

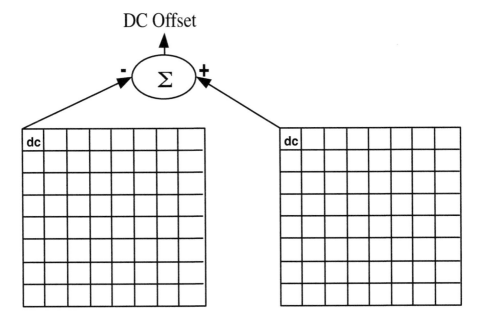

Figure 1.6: DC offset.

### 1.4.1.5 Encoding AC Coefficients

AC coefficients are encoded independently within a block. It has been observed that, after quantization, many of these coefficients turn out to be almost zero values. In particular, high-frequency components are more prone to assuming zero values. In 2-D, ordering of spatial frequencies does not follow exactly the row-wise or column-wise raster scan. One has to follow a zigzag scanning starting from the leftmost and topmost location and moving diagonally after each horizontal or vertical shift (as shown in Figure 1.7). This converts a 2-D sequence into an 1-D sequence, where trailing members are always of higher-order spatial frequency. This sequence is further converted into runs of zeroes terminated by a nonzero element. Finally, all the trailing zeroes are terminated by an *end of block* symbol. *AC coefficients* are represented in this way. At a later stage, these runs (number of zeroes followed by a nonzero coefficient) are efficiently encoded by following the Huffman encoding or arithmetic encoding scheme.

### 1.4.1.6 Entropy Encoding

There are recommended Huffman encoding tables for encoding both DC and AC coefficients. However, we can also define a new encoding table or adopt arithmetic encoding in producing the final compressed stream. Huffman codes are self-terminating codes, and the encoding technique tries to minimize the

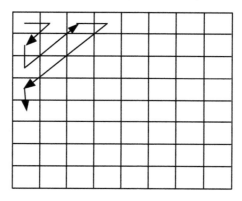

Figure 1.7: Zigzag sequence of AC coefficients.

number of bits required for representing a set of symbols whose information content is measured by its *entropy*. The encoding scheme adopts a strategy of assigning variable length codes to these symbols such that less probable symbols are represented by longer encoded binary streams. By looking at the statistical distribution of intermediate symbols, these codes are generated. Huffman tables for DC coefficients and runs of AC coefficients are different. However, for encoding AC coefficients, JPEG uses both the tables.

### 1.4.1.7　Encoding Colors

As color components in the RGB color space are significantly correlated, in JPEG compression standard, color images are represented in the YCbCr color space [52]. The YCbCr color space is related to the RGB color space as follows

$$
\begin{aligned}
Y &= 0.502G + 0.098B + 0.256R, \\
Cb &= -0.290G + 0.438B - 0.148R + 128, \\
Cr &= -0.366G - 0.071B + 0.438R + 128.
\end{aligned} \tag{1.6}
$$

In the above transformation, it is assumed that pixels of R, G, and B components are represented by 8 bits. In this space, $Y$ is referred to as the *luminance component*, while $Cb$ and $Cr$ are referred to as *chrominance components*. The $Cb$-$Cr$ components have less bandwidth than their luminance counterpart. That is why they can be further downsampled leading to a significant reduction in the data size. In the JPEG standard, these three color components are encoded in various proportions. With $1 : 1 : 1$, $Cb$ and $Cr$ components are compressed with full resolution, signifying that every luminance block has its corresponding chrominance block. On the other hand, $4 : 1 : 1$ imply that, for every 4 luminance ($Y$) blocks there is only *one Cb* and *one Cr* block. This means that $Cb$ and $Cr$ are *halfsampled* along both the spatial directions. In fact, the latter scheme is more popular while compressing color images. A group of such blocks (corresponding luminance and chrominance blocks) is referred to as a *macroblock*, which forms a unit for the coding structure of the compressed stream (see Figure 1.8). It may also be noted that the *quantization tables* for chrominance components are different from that of the luminance component.

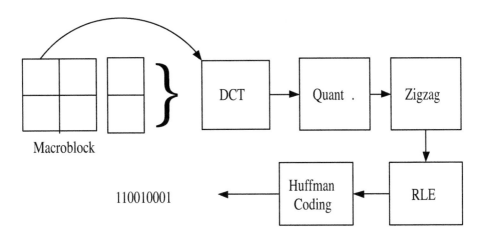

Figure 1.8: Encoding steps of a macroblock.

## 1.4.2 JPEG2000

It has been mentioned earlier that JPEG adopted another image compression standard, JPEG2000 [27, 119], which is based on DWT [4]. Usually, JPEG2000 is expected to have a higher compression ratio given the same level of distortion compared to the DCT-based JPEG compression standard. There are also other advantages to JPEG2000 in terms of scalability, encoding regions of interests with varying level of compression, progressive reconstruction of images from the same compressed stream, same computational framework for lossy and lossless schemes, etc. For incorporating these features, the JPEG2000 code-structure is more complex than that of the earlier JPEG scheme, thus introducing more complexity in its encoder and decoder design. JPEG2000 is also capable of handling images of larger dynamic ranges, while in JPEG it is assumed that a component should have pixel values between 0 and 255 (8-bit representation of images).

In JPEG2000, images are partitioned into *tiles* of larger size, and each individual *tile* is encoded independently. Even the whole image could form a single tile. Different stages in the process of compressing an image are shown in Figure 1.9. The same schema holds for both lossless and lossy compression. In the following text, these operations are reviewed briefly.

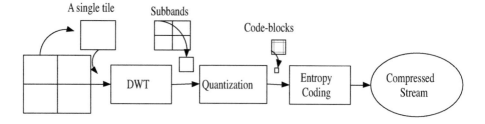

Figure 1.9: Steps of JPEG2000 compression technique.

### 1.4.2.1  Discrete Wavelet Transform (DWT)

The DWT is applied to every nonoverlapping tile of an image, resulting in its decomposition into several subbands. Before this, the tile goes through a level-shifting operation depending upon the number of bits (say, $n$) used for representing a pixel. The level-shifting operation brings pixel values between 0 and $2^n - 1$ to $-2^{n-1}$ and $2^{n-1} - 1$.

The DWT is applied row-wise and column-wise separately for decomposing the image into *four* subbands. Each subband is obtained by applying a combination of low-pass and/or high-pass filters in successive direction. These filters are known as *analysis filters*. As the bandwidth of the filtered images are reduced compared to the original image, they are downsampled or subsampled by a factor of two so that the sum total of the number of samples in all the subbands remains the same. During decoding, these subsampled subbands are upsampled (or interpolated to full resolution), and then they are subjected to another set of combination of low-pass and/or high-pass filters. These filters are known as *synthesis filters*. Finally, the sum of all these filtered responses produces the reconstructed image. In Figure 1.10, a schematic diagram is presented showing the computation of the DWT as well as its inverse in 1-D.

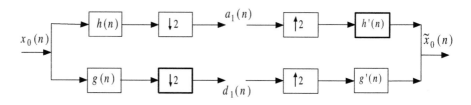

**Figure 1.10**: Discrete Wavelet Transform and its inverse in 1-D. (From [147] with permission from the author.)

In Figure 1.10, the signal $x_0(n)$ is transformed into wavelet coefficients in

**Table 1.1**: Daubechies 9/7 analysis and synthesis filter banks for lossy compression

|  | Analysis Filter Bank | | Synthesis Filter Bank | |
|---|---|---|---|---|
| $n$ | $h(n)$ | $g(n-1)$ | $h'(n)$ | $g'(n+1)$ |
| 0 | 0.603 | 1.115 | 1.115 | 0.603 |
| $\pm 1$ | 0.267 | $-0.591$ | 0.591 | $-0.267$ |
| $\pm 2$ | $-0.078$ | $-0.058$ | $-0.058$ | $-0.078$ |
| $\pm 3$ | $-0.017$ | 0.091 | $-0.091$ | 0.017 |
| $\pm 4$ | 0.027 | | | 0.027 |

Table 1.2: 5/3 analysis and synthesis filter banks for lossless compression

|  | Analysis Filter Bank | | Synthesis Filter Bank | |
|---|---|---|---|---|
| $n$ | $h(n)$ | $g(n-1)$ | $h'(n)$ | $g'(n+1)$ |
| 0 | $\frac{6}{8}$ | 1 | 1 | $\frac{6}{8}$ |
| $\pm 1$ | $\frac{2}{8}$ | $-\frac{1}{2}$ | $\frac{1}{2}$ | $-\frac{2}{8}$ |
| $\pm 2$ | $-\frac{1}{8}$ | | | $-\frac{1}{8}$ |

the form of $a_1(n)$ and $d_1(n)$, known as *approximation* and *detail* coefficients, respectively. Corresponding low-pass and high-pass FIR (Finite Impulse Response) filters [100] are $h(n)$ and $g(n)$. They form the *analysis filter bank*. In the decoding stage, respective synthesis filters are $h'(n)$ and $g'(n)$. Following quadrature mirror filtering (QMF) [100], we may design these four filter banks (including both analysis and synthesis filters) such that perfect reconstruction is possible (i.e., according to Figure 1.10, $\tilde{x}_0(n) = x_0(n)$). In JPEG2000, two different sets of filter banks are used for lossy and lossless compression. They are shown in Tables 1.1 [4] and 1.2 [48]. For lossless compression, we need to represent wavelet coefficients with integers only so as to avoid errors due to impreciseness of floating-point representation. Hence, corresponding filters are implemented using only integer arithmetic, where the set of valid operations include addition, multiplication, and shift (for performing division or multiplication) with integers as operands. For this reason, this class of discrete wavelet transform is also termed as *integer wavelet transform* (IWT) [34]. During filtration, boundaries are *symmetrically and periodically* extended at both ends to avoid discontinuity at those sample points of a signal. However, the number of additional samples that are required for this periodic and symmetric extension depends on the length of the FIR filter.

For an 1-D sequence, the coefficients are usually arranged in such a fashion that $d_1(n)$ follows $a_1(n)$. This kind of decomposition of signal is referred to as *dyadic decomposition* [92]. The low-pass filtered subband may be subjected to another level of dyadic decomposition in the same way, and this may continue recursively as long as the resolution of the low-pass subband at that level permits. In JPEG2000, images are decomposed into *four* subbands by filtering with *low–low*, *low–high*, *high–low* and *high–high* filters. These subbands are referred as *LL, LH, HL,* and *HH,* respectively. A single-level *dyadic decomposition* is shown in Figure 1.11(a). The *LL* band may further be decomposed in the same way. An example of 3-level decomposition is shown in Figure 1.11(b). In this case the corresponding subbands are shown as $LL_3$, $LH_3$, $HL_3$, $HH_3$, $LH_2$, $HL_2$, $HH_2$,$LH_1$, $HL_1$, and $HH_1$. These subbands are separately subjected to a scalar quantization, as discussed below.

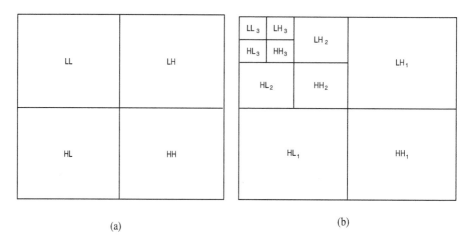

(a)                                   (b)

Figure 1.11: Dyadic decomposition of images: (a) single-level and (b) 3-level.

#### 1.4.2.2   Quantization

In JPEG2000, each subband is independently quantized. However, unlike JPEG, all the coefficients of a subband use the same quantization threshold. The quantization step ($\Delta$) is expressed with respect to its dynamic range as follows:

$$\Delta = 2^{n-\epsilon}(1 + \frac{\mu}{2^{11}}). \qquad (1.7)$$

In the above equation, $n$ is considered as the *nominal dynamic range* of the subband. It is the maximum number of bits required to represent the subband, which is the sum of the number of pixel bits in the original tile and additional bits required for representing the values after the analysis. For example, the value of $n$ for the $HH_1$ subband of an image represented by 8 bits per pixel is 10. Parameters $\epsilon$ and $\mu$ are the number of bits allotted to the *exponent* and *mantissa* respectively, of its coefficients. Given $X(u, v)$ as the coefficients of a subband, the quantized values are obtained as follows:

$$X_q(u, v) = sign(X(u, v)) \lfloor \frac{|X(u, v)|}{\Delta} \rfloor. \qquad (1.8)$$

For lossless compression, $\Delta$ is 1, implying that values of $\epsilon$ and $\mu$ are $n$ and 0, respectively. There is no recommended set of $\mu$ and $\epsilon$ for determining the quantization step for a subband, and this may be *explicitly* assigned for each of them. In that case, these parameters are to be included in the encoding stream. However, in an *implicit quantization*, given those parameters for the lowest level of resolution (e.g. the corresponding to $LL$ band at level $k$) as $\mu_0$ and $\epsilon_0$, values for the $i$th subbands are determined by the following expressions:

$$\begin{aligned} \mu_i &= \mu_0, \\ \epsilon_i &= \epsilon_0 + i - k. \end{aligned} \qquad (1.9)$$

### 1.4.2.3 Bit-Stream Layering, Packetization, and Entropy Coding

In the final stage of an encoder, a compressed bit stream is produced so that the decoder is capable of progressively decoding a region of interest with a desired resolution. In its code structure, every subband is partitioned into a set of nonoverlapping code blocks. Each code block is independently encoded by a scheme known as *embedded block coding on truncation* (EBCOT) [142]. In this scheme, each bit-plane of wavelet coefficients is processed by three passes, namely, *significant propagation*, *magnitude refinement*, and *clean-up*. The resulting bit-streams are encoded using *arithmetic coding*. A *layer* is formed with the output of similar passes from a group of code blocks. In a *layer*, *packets* are formed by grouping *corresponding code blocks* of subbands at the same level of decomposition. They are also known as *precincts* (see Figure 1.12).

### 1.4.2.4 Color Encoding

As with JPEG, in JPEG2000 also color components are transformed into a different space from the original RGB space to reduce redundancy among the components. However, unlike JPEG, in this case there is no requirement of downsampling of chrominance components on account of the multiresolution representation used in JPEG2000. For lossy compression, the irreversible component transform is used as follows (see Eq.(1.10)):

$$\begin{aligned} Y &= 0.587G + 0.114B + 0.299R, \\ Cb &= -0.33126G + 0.5B - 0.16875R, \\ Cr &= -0.41869G - 0.08131B + 0.5R. \end{aligned} \qquad (1.10)$$

The inverse transform of the above is given by

$$\begin{aligned} R &= Y + 1.402Cr, \\ G &= Y - 0.34413Cb + 0.71414Cr, \\ B &= Y + 1.772Cb. \end{aligned} \qquad (1.11)$$

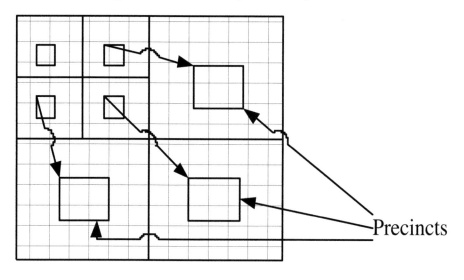

Figure 1.12: Code structure of JPEG2000.

For lossless compression, reversible color transform is applied as follows:

$$\begin{aligned} Y &= \lfloor \tfrac{R+2G+b}{4} \rfloor, \\ U &= R - G, \\ V &= B - G. \end{aligned} \qquad (1.12)$$

The corresponding inverse of the reversible color transformation is given by

$$\begin{aligned} G &= Y - \lfloor \tfrac{U+V}{4} \rfloor, \\ R &= U + G, \\ B &= V + G. \end{aligned} \qquad (1.13)$$

After obtaining color components by applying one of the above transforms, individual components are independently compressed following the same steps earlier described.

## 1.5   Video Compression Techniques

There exists greater amount of statistical redundancies in video data compared to still images. In addition to spatial redundancy, it also has temporal redundancy in consecutive frames, and compression algorithms aim at discarding these redundancies, leading to reduction in the size of the video. Usually, video compression is lossy in nature as it deals with enormous amount of data. The block diagram of a typical video encoder is shown in Figure 1.13.

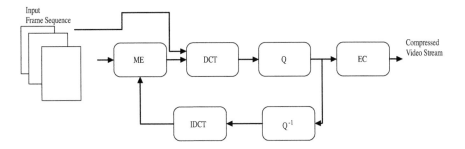

Figure 1.13: Block diagram of a typical video encoder.

In a video compression algorithm, each input video frame is compressed individually. First, they are mapped from the RGB color-space to the YCbCr color-space. Like the baseline lossy JPEG compression algorithm, usually the luminance component, $Y$, of the frame is sampled at the original picture resolution, while the chrominance components, $Cb$ and $Cr$, are downsampled by two in both the horizontal and vertical directions to yield a 4:2:0 subsampled picture format. Input frames are further subdivided into macroblocks (MBs) of typical size of $16 \times 16$ pixels. Hence, each YCbCr macroblock consists of four luminance blocks of dimensions $8 \times 8$ pixels, followed by one $8 \times 8$ $Cb$ block and one $8 \times 8$ $Cr$ block. As shown in Figure 1.13, an input video frame first undergoes motion estimation (ME) with respect to one or more reference frames. Some frames (or some macroblocks of a frame) are independently encoded, and the ME block is skipped for them. The motion estimated frame is then transformed from spatial to frequency domain using an $8 \times 8$ block employing DCT. The transformed coefficients are then quantized (Q) and entropy coded (EC) using variable length codes (VLC) to generate the output bitstream. Again, on the encoder side, the quantized coefficients are dequantized ($Q^{-1}$) and inverse transformed (IDCT) to obtain the reconstructed frame. The reconstructed frame is then subsequently used to estimate the motion in the next input frame.

Different video compression algorithms have been developed for different types of applications in keeping with various targets set by them. For example, H.26x standards are targeted at video teleconferencing, while the Motion JPEG (MJPEG) is convenient for video editors as it has easy access to individual frames. However, the latter technique works with a low compression ratio. There are different standards suggested by the MPEG, starting from MPEG-1, which is designed for low bitrate (e.g., 1.5 Mbps) audio and video playback applications, and then MPEG-2 for higher bitrates for high-quality playback applications with full-sized images (e.g., CCIR 601 with studio quality at 4–10 Mbps). Later, MPEG-4 video standard was advanced for both low

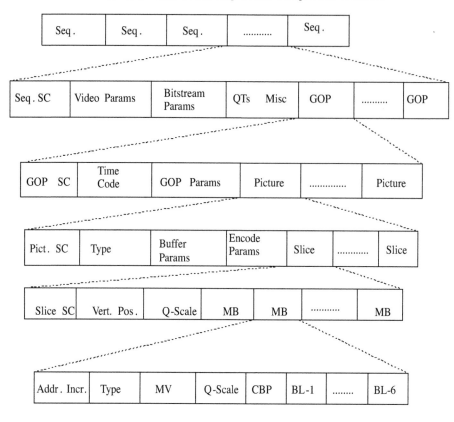

Figure 1.14: Video stream data hierarchy.

data rate and high data rate applications. Under the context of compressed domain processing, let us review three such video compression standards, namely MPEG-2, MPEG-4, and H.264 for their extensive use in different applications.

## 1.5.1  MPEG-2

The MPEG-2 [135] is one of the most popular video compression techniques. The structure of the encoded video stream and its encoding methodology are briefly described below.

### 1.5.1.1  Encoding Structure

Considering a video as a sequence of frames (or pictures), in MPEG-2 video stream, the video data is organized in a hierarchical fashion as shown in Figure 1.14. The video stream consists of five layers : GOP, pictures, slices, macroblock, and block, as discussed in the following text (see Figure 1.15).

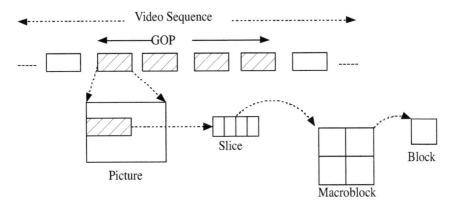

Figure 1.15: Video sequence in an MPEG stream.

1. **Video Sequence:** It begins with a sequence header, includes one or more groups of pictures, and ends with an end-of-sequence code.

2. **Group of Pictures (GOP):** This consists of a header and a series of one or more pictures intended to allow random access into the sequence.

3. **Picture:** A *picture* is an individual image or frame, the primary coding unit of a video sequence. A frame of a colored video has three components: a luminance ($Y$) and two chrominance ($Cb$ and $Cr$) components. The $Cb$ and $Cr$ components are one half the size of the $Y$ in horizontal and vertical directions.

4. **Slice:** One or more contiguous macroblocks define a *slice*. In a slice, macroblocks are ordered from left to right and top to bottom. Slices are used for handling of errors. If an error is reported for a slice, a decoder may skip the erroneous part and goes to the start of the next slice.

5. **Macroblock:** Like still image coding, the basic coding unit in the MPEG algorithm is also a *macroblock*. It is a $16 \times 16$ pixel segment in a frame. If each chrominance component has one-half the vertical and horizontal resolution of the luminance component, a macroblock consists of four $Y$, one $Cr$, and one $Cb$ block (as shown in Figure 1.16).

6. **Block:** A *block* is the smallest coding unit in the MPEG-2 algorithm. It is of $8 \times 8$ pixels, and it can be one of three types: luminance ($Y$), complementary red chrominance ($Cr$), or complementary blue chrominance ($Cb$).

### 1.5.1.2 Frame Types

There are three types of pictures or frames in the MPEG standard, depending upon their role in the encoding process.

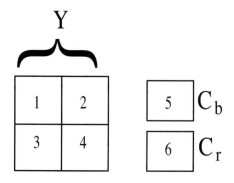

Figure 1.16: Structure of a macroblock.

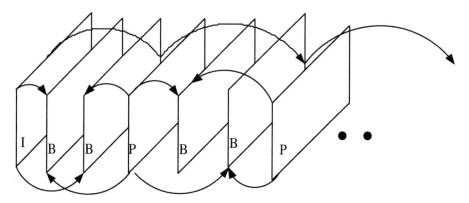

Figure 1.17: Prediction of pictures in a GOP sequence.

1. Intra-frames (I-Frames)

2. Predicted frames (P-Frames)

3. Bidirectional frames (B-Frames)

A sequence of these three types of pictures forms a group of pictures whose starting frame must be of an 'I-frame.' A typical GOP structure is shown in Figure 1.17.

1. **Intra-frames:** Intra-frames, or I-frames, are coded independently using only information present in the picture itself. It enables one to synchronize a video with potential random access points within its compressed stream. Encoding of I-frames is the same as still-image encoding. In MPEG-2, the same JPEG baseline lossy compression algorithm is used for coding these frames.

2. **Predicted frames:** A predicted frame, or P-frame, is predicted from its nearest previous I- or P-frame. The prediction in this case is guided by motion compensation, leading to higher compression of these frames.

3. **Bidirectional frames:** Bidirectional frames, or B-frames, are frames that use both a past and a future frame as references. Hence, in this case, the frame is encoded with bidirectional prediction. Expectedly, B-frames have higher compression than P-frames. However, it takes more computation during prediction.

### 1.5.1.3   Method of Encoding Pictures

1. **Intra-frame:** As mentioned earlier, I-frames are encoded following the same baseline JPEG lossy compression algorithm as discussed earlier. In this case also, a color component of an I-frame is partitioned into a set of $8 \times 8$ nonoverlapping blocks, and each partition is subjected to operations such as level shifting, forward DCT, quantization, and entropy encoding of DC and AC coefficients, which are organized in a zigzag order to produce long runs of zero, encoded subsequently with a variable length Huffman code (see Figure 1.8).

2. **P-frame:** A P-frame is coded with reference to a previous image (reference image) that is either an I- or a P-frame as shown in Figure 1.18. In this case, the frame is partitioned into a set of $16 \times 16$ blocks (called macroblocks under this context), and each of them is predicted from a macroblock of reference frame (of the same size). The respective reference macroblock is obtained by searching in the reference image around the neighborhood of its same position. The offset position of the reference macroblock, which provides minimum error (usually computed in the form of the sum of absolute differences (SAD)), is stored in the compressed stream, and it is known as the *motion vector* of the corresponding macroblock of the P-frame. Finally, the difference between the values of the encoded macroblock and reference macroblock (i.e., the prediction error) is computed, and they are encoded by subsequent application of *forward DCT, quantization, run-length encoding of zigzag sequence of AC coefficients*, and *Huffman encoding of respective runs and DC off-sets*.

3. **B-frame:** A B-frame is bidirectionally predicted from a past and a future reference frame. In this case, motion vectors can be from either the previous reference frame, or from the next picture, or from both. This process of encoding is shown in Figure 1.19. Consider a B-frame **B**, predicted from two reference frames $R_1$ and $R_2$. Let $R_1$ be the past I- or P-frame, and $R_2$ be the future I- or P- frame. For each macroblock $m_B$ of **B**, the closest match $m_1$ in $R_1$ and $m_2$ in $R_2$ are computed. In that case, the prediction of $m_b$, $\hat{m}_b$, is obtained as follows.

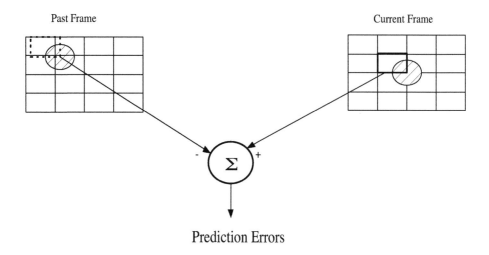

Figure 1.18: Coding of predicted pictures.

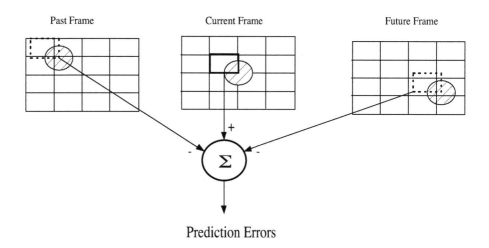

Figure 1.19: Coding of bidirectional predicted pictures.

$$\hat{m}_b = round(\alpha_1 m_1 + \alpha_2 m_2),$$

where $\alpha_1$ and $\alpha_2$ are defined below.

(a) $\alpha_1 = 0.5$, and $\alpha_2 = 0.5$ if both matches are satisfactory.

(b) $\alpha_1 = 1$, and $\alpha_2 = 0$ if only first match is satisfactory.

(c) $\alpha_1 = 0$, and $\alpha_2 = 1$ if only second match is satisfactory.

(d) $\alpha_1 = 0$, and $\alpha_2 = 0$ if neither match is satisfactory.

Finally, the error block $e_b$ is computed by taking the difference of $m_b$ and $\hat{m}_b$. These error blocks are coded in the same way as the blocks of an I-frame.

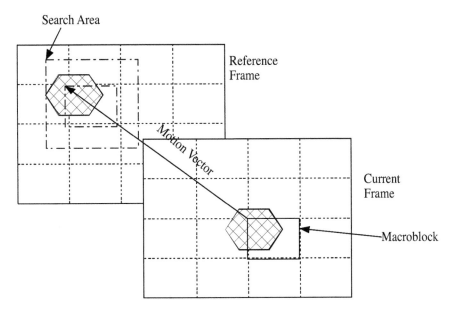

Figure 1.20: Motion estimation for MPEG-2 video encoder.

#### 1.5.1.4 Motion Estimation

In video compression, for motion-compensated prediction, pixels within the current frame are modeled as translations of those within a reference frame. In forward prediction, each macroblock (MB) is predicted from the previous frame assuming that all the pixels within that MB undergo same amount of translational motion. This motion information is represented by a two-dimensional displacement vector or *motion vector*. Due to its block-based representation, block-matching techniques are employed for motion estimation (see Figure 1.20). As shown in the figure, both the current frame and the reference frame have been divided into blocks. Subsequently, each block in the current frame is matched at all locations within the search window of the previous frame. In this block-based matching technique, a cost function measuring the mismatch between a current MB and the reference MB is minimized to provide the motion vector. There are different cost measures used for this purpose, such as, mean of absolute differences (MAD), sum-of-absolute-differences (SAD), mean-square-error (MSE), etc. The most widely used metric is the SAD, defined by

$$SAD_{i,j}(u,v) = \sum_{p=0}^{N-1} \sum_{q=0}^{N-1} |c_{i,j}(p,q) - r_{i-u,j-v}(p,q)|. \qquad (1.14)$$

where $SAD_{i,j}(u,v)$ represents the SAD between the $(i,j)th$ block and the block at the $(u,v)th$ location in the search window $W_{i,j}$ of the $(i,j)th$ block. Here, $c_{i,j}(p,q)$ represents the $(p,q)th$ pixel of an $N \times N$ $(i,j)th$ MB $C_{i,j}$, from the current picture. $r_{i-u,j-v}(p,q)$ represents the $(p,q)th$ pixel of an $N \times N$ MB from the reference picture displaced by the vector $(u,v)$ within the search range of $C_{i,j}$. To find the MB producing the minimum mismatch error, the SAD is to be computed at several locations within the search window. The simplest but the most computationally intensive search method, known as the full search or exhaustive search method, evaluates SAD at every possible pixel location in the search area. Using full search, the motion vector is computed as follows:

$$MV_{i,j} = \{(u',v') | SAD_{i,j}(u',v') \le SAD_{i,j}(u,v), \forall (u,v) \in W_{i,j}\}, \qquad (1.15)$$

where $MV_{i,j}$ expresses the motion vector of the current block $C_{i,j}$ with minimum SAD among all search positions. In MPEG-2, motion vectors are computed either with *full pixel* or *half pixel* precision. In the former case, MBs are defined from the locations of the reference frame in its original resolution, but for *half pixel* motion vectors, the reference image is first bilinearly extrapolated to double its resolution in both the directions. Then motion vectors are computed from the locations of the interpolated reference image. For a downsampled chrominance component, the same motion vector is used for prediction. In this case, the resulting motion vector of the chrominance MB is scaled down by a factor of two.

### 1.5.1.5   Handling Interlaced Video

The MPEG-2 compression standard also handles interlaced video, which is common for television standards. In this case, a frame is partitioned into two *fields* (*odd* and *even* fields). Each *field* is separately encoded, and motion estimation of an MB of a *field* is optionally performed from *the same type of field* of the reference frame or *another field* of the current frame if it is encoded prior to the present one.

## 1.5.2   MPEG-4

The MPEG-4 [135] video compression technique is distinguished by the fact that it takes care of object-based encoding of a video. The project was initiated by MPEG in July 1993 in view of its application in representing multimedia content and delivery. The standard was finally adopted in February 1999. In fact, the standard encompasses representation of not only video but also other medias such as synthetic scenes, audio, text, and graphics. All these entities are encapsulated in an *audiovisual object* (AVO). A set of AVOs represents a multimedia content, and they are composed together to form the final compound AVOs or scenes. In our discussion, we restrict ourselves to the video compression part of the MPEG-4 standard. An overview of the video compression technique is shown in Figure 1.21.

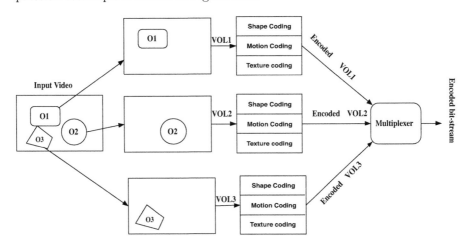

Figure 1.21: An overview of an MPEG-4 video encoder.

### 1.5.2.1   Video Object Layer

As shown in Figure 1.21, an input video is modeled as a set of sequence of video objects. We have to apply segmentation algorithms for extracting objects from individual frames known as *video object planes* (VOPs), and the sequence of these objects in the overall video defines a *video object* (VO) or a *video object layer* (VOL). For example, in Figure 1.21, there are three such VOLs. We may also consider the *background* as another layer. However, in MPEG-4, the background of a video may be efficiently coded as *sprites*, which is discussed later. Individual VOs are independently coded consisting of information related to *shape*, *motion*, and *texture*. Each of these codings is briefly discussed here.

1. **Shape encoding:** Every VOP in a frame could be of arbitrary shape. However, the bounding rectangle of this shape in the frame is specified.

This rectangle is adjusted in such a way that its dimension in both horizontal and vertical directions becomes an integral multiple of 16 so that it can be encoded as a set of nonoverlapping MBs of size 16 × 16. The pixels within the rectangle not belonging to the object are usually denoted by the value zero (0); otherwise they contain 1 (in binary representation of the shape) or a gray value (usually denoted by $\alpha$), used for blending with other VOPs during reconstruction of the video. There are three types of MBs within this rectangle. An MB may have (i) all nonzero pixels (contained within the object), or (ii) all zero pixels, or (iii) both types of pixels. The third type of MB is called the *boundary* MB of a VOP. The binary shape information of a VOP is encoded by a *content-adaptive arithmetic coding* (CAE) [131], while for gray-shape representation the usual *motion compensated* (MC) DCT representation is followed.

2. **Motion encoding:** Like MPEG-2, each VOP of a layer is one of three types, namely, *I-VOP*, *P-VOP*, and *B-VOP*. In this case also, motion vectors are computed in the same manner as done for MPEG-2. However, there are other options (or modes) for computing motion vectors and subsequently obtaining the motion-compensated prediction errors for an MB. These modes include the following:

   (a) Four motion vectors for each 8 × 8 block of an MB are computed separately, and the prediction errors are obtained using them.

   (b) Three overlapping blocks of the reference frame are used to compute the prediction errors for each 8 × 8 block of the current MB. These reference blocks are obtained from the closest matches of its neighboring blocks, either to its left (right) or to its top (bottom). The third one is the closest match of the concerned block itself in the reference frame. For every pixel in the current block, a weighted mean of corresponding pixels of those three blocks provides its prediction. These weights are also predefined in the standard.

3. **Texture encoding:** In texture coding, MBs of a VOP are encoded in the same way as it is done in MPEG-2. For I-VOP, 8 × 8 blocks are transformed by the DCT, and subsequently transformed coefficients are quantized and entropy coded. For P-VOP and B-VOP, after motion compensation of an MB, residual errors (for each 8 × 8 block) are encoded in the same way. However, there are a few additional features in MPEG-4 in encoding textures. They are briefly discussed here.

   (a) **Intra DC and AC prediction:** In MPEG-4 there is a provision for prediction of *quantized* DC and AC coefficients from one of its neighboring blocks (either to its left or to its bottom) in intra VOPs. The neighboring block, which has lower gradient in the coefficient space (with respect to the leftmost and topmost diagonal neighboring block), is chosen for this purpose. The same feature is also extended for the MBs in the intra mode of inter VOPs.

(b) **DCT of boundary blocks:** As boundary blocks have pixels not belonging to a VOP, they need to be either padded with suitable values to avoid abrupt transitions in the block (as they may demand greater number of bits for encoding), or they may be ignored in the computation of the transform itself. In the first strategy, all the non-VOP pixels within the block are padded with the mean pixel value of that block, and then they are subjected to a low-pass average filtering to reduce the abrupt transitions near the junctions of non-VOP and VOP pixels. Alternatively, for dropping these non-VOP pixels from the compressed stream, we need to apply *shape adaptive DCT (SA-DCT)* [131] to the corresponding block. In this case, $N$-point DCTs are applied successively to the columns and rows of the block containing $N$ ($N \leq 8$) VOP pixels.

### 1.5.2.2   Background Encoding

The background of a video may be treated as a separate VOL and encoded in the same way as others. However, MPEG-4 also makes provision for encoding in the form of a sprite. A sprite is a still image formed from a set of consecutive video frames, that do not show any motion within its pixel. At any instance of time, a particular rectangular zone of the sprite is covered by a frame. Hence, once a sprite is defined for the set of frames, it is sufficient to pass the coordinate information of that zone for rendering the concerned frame. This is similar to the *panning* of the camera, and the respective parameters are encoded in the video stream. A sprite acts as a background for that set of frames. If it is defined for the complete video, the sprite is *static*, and it is computed *offline* during the encoding process. However, in real time, we have to compute *dynamic* sprites for a group of pictures.

### 1.5.2.3   Wavelet Encoding of Still Images

The MPEG-4 allows use of wavelets for encoding of still images. Like JPEG2000, it also follows *dyadic decomposition* [92] of images into a multi-resolution representation. However, the coefficients of the $LL$ band (see Section 1.4.2.1 and Figure 1.11) is encoded in a different way from other subbands. In this case, they are simply quantized and entropy coded by *arithmetic coding*. Other subbands are encoded with the help of a *zero tree* as proposed in [128]. The symbols of the zero tree are encoded using an *arithmetic coder*.

### 1.5.3   H.264/AVC

The video compression standard H.264/AVC [155] (or H.264 as will be referred to subsequently) is the outcome of the joint effort of the ITU-T VCEG and the ISO/IEC MPEG. The major objectives of this standardization effort are (i) to develop a simple and straightforward video coding design with enhanced compression performance, and (ii) to provide a network-friendly video representation addressing both conversational (such as video telephony) and non-conversational (such as storage, broadcast, or streaming) applications. In fact, the H.264 has greatly outperformed other existing video standards such as MPEG-2.

The H.264 specifies the video in the form of a *video-coding layer* (VCL) and a *network abstraction layer* (NAL) (see Figure 1.22). The VCL represents the video content, and the NAL formats the VCL with requisite header information appropriate for its transmission by transport layers or for its storage in a media. The H.264 video compression algorithm is also a hybrid of inter-picture prediction and transform coding of the prediction residual error (see Figure 1.23).

Like MPEG-2 and MPEG-4, the H.264 design [155] also supports the coding of video in 4:2:0 chroma format. It handles both progressive and interlaced videos. In Figure 1.23, processing of a single MB of a video frame is shown with the help of a block diagram. The basic coding structure of H.264 is a hierarchical one [151]. A video sequence consists of several *pictures* or *frames*. A single picture is partitioned into a set of *slices*, which is again a collection of MBs. Some of the features of H.264 and its coding structure are discussed next.

### 1.5.3.1  Slices and Slice Groups

A video frame may be split into one or several slices. The H.264 has the provision of defining a slice with a *flexible macroblock ordering (FMO)* by utilizing the concept of a *slice group*, as a set of MBs and specified by a mapping between an MB to a slice. Again, each slice group consists of one or more slices such that MBs in the slice are processed in the order of a raster scan.

### 1.5.3.2  Additional Picture Types

There are two more additional types for encoding slices other than usual picture types such as I, P and B. These two new coding types are switching I (SI) and switching P (SP) pictures, which are used for switching the bitstream from one rate to another [75].

### 1.5.3.3  Adaptive Frame/Field-Coding Operation

For higher coding efficiency, the H.264/AVC encoders may adaptively encode fields (in the interlaced mode) separately or combine them into a single frame.

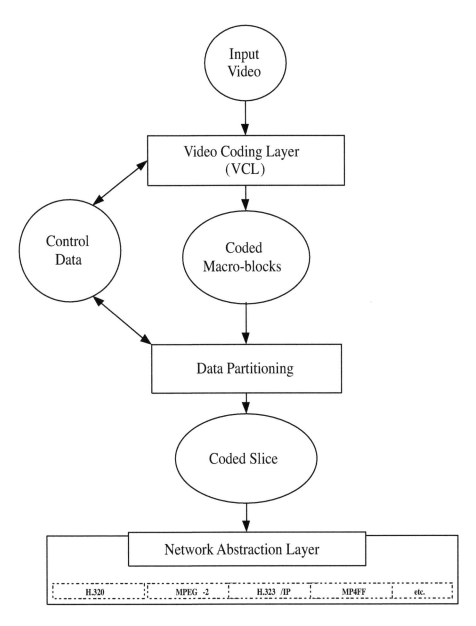

Figure 1.22: Structure of H.264/AVC video encoder.

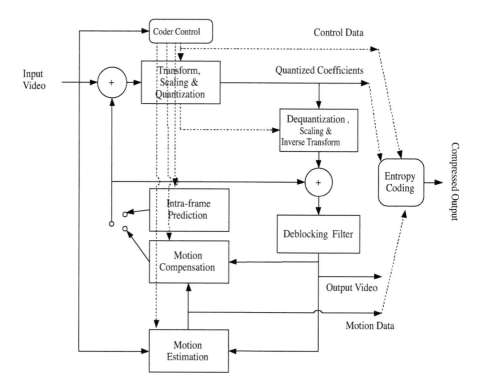

Figure 1.23: Basic coding structure of H.264/AVC.

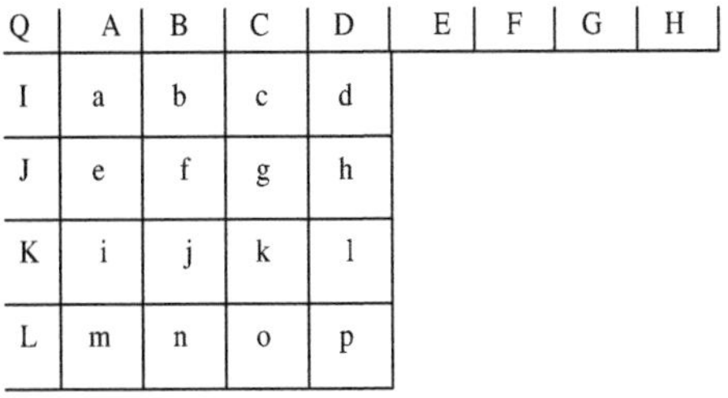

| Q | A | B | C | D | E | F | G | H |
|---|---|---|---|---|---|---|---|---|
| I | a | b | c | d | | | | |
| J | e | f | g | h | | | | |
| K | i | j | k | l | | | | |
| L | m | n | o | p | | | | |

**Figure 1.24**: Intra_4 × 4 prediction is computed for samples $a - p$ of a block using samples $A - Q$.

#### 1.5.3.4 Intra-frame Prediction

There are different modes of intra-frame predictions supported in H.264, namely, $Intra\_4 \times 4$ or $Intra\_16 \times 16$, and $I\_PCM$ prediction modes. These are discussed below.

1. I_PCM: This coding type allows the encoder to simply bypass the prediction.

2. Intra_4 × 4 mode: In this mode, each 4 block is predicted from spatially neighboring samples as illustrated in Figure 1.24. The 16 samples of the 4 × 4 block as labeled from $a$ to $p$ are predicted from decoded samples of adjacent blocks (labeled by big alphabets from $A$ to $Q$). For each 4 × 4 block, there may be one of nine prediction modes (see Figure 1.25).

   In *mode 0 (vertical prediction)*, samples above the 4 × 4 block are simply copied into the block (as indicated by the arrows). Similarly in *mode 1 (horizontal prediction)*, the samples to the left of the 4 × 4 block are copied. In *mode 2 (DC prediction)*, averages of adjacent samples are used for prediction. The remaining six modes are *diagonal prediction* modes, namely, *diagonal-down-left*, *diagonal-down-right*, *vertical-right*, *horizontal-down*, *vertical-left*, and *horizontal-up* prediction. Pixels along those specified directions are used for prediction.

3. Intra_16 × 16 mode: In this mode the complete luminance component of an MB is predicted from neighboring MBs. It has four prediction modes, namely, *mode 0* (vertical prediction), *mode 1* (horizontal prediction), *mode 2* (DC prediction) and *mode 3* (plane prediction). The first three modes are the same as the Intra_4 × 4 mode, except for the fact that, instead of four neighbors on each side to predict a 4 × 4 block, 16 neighbors on each side are used to predict the 16 × 16 block. In *mode 3* (plane prediction), predictions are made from both vertical and horizontal neighbors lying on the same straight line parallel to the off-diagonal.

   The chrominance samples of an MB usually have less variation over a

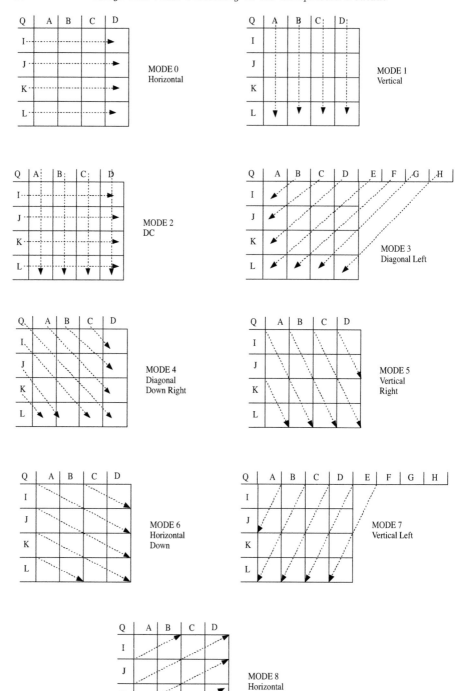

Figure 1.25: Intra $4 \times 4$ prediction modes.

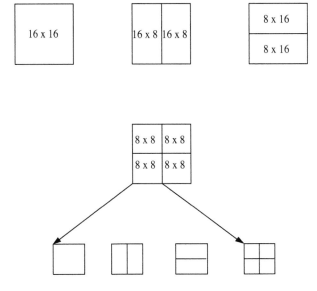

Figure 1.26: Partitioning of an MB.

relatively larger area. Hence they are predicted using a similar technique as used for the luminance component in Intra-16 × 16 mode.

### 1.5.3.5 Inter-frame Prediction in P Slices

There are different types of partitioning in a *P-MB* for motion compensation. For example, a 16 × 16 luminance block may be partitioned as blocks of one 16 × 16, two 16 × 8 or 8 × 16, and four 8 × 8 blocks. Again, if an 8 × 8 partition is chosen, it may be further decomposed into partitions of 8 × 4, 4 × 8, or 4 × 4 blocks. In Figure 1.26, different partitioning schemes are illustrated.

As can be seen from different partitioning schemes, a maximum of 16 motion vectors may be transmitted for a single P MB. The accuracy of motion compensation may be in units of one, half, or one quarter of the distance between luminance samples. For full pixel motion vectors, integer-coordinate locations within the frame itself are specified. However, for half or one-quarter pixel vectors, the frame is interpolated to generate noninteger positions. Samples at half-sample positions are predicted by applying a one-dimensional 6-tap FIR filter horizontally and vertically, whereas at quarter-sample positions these are generated by averaging samples at integer- and half-sample positions [155]. The samples for the chrominance components are always predicted by bilinear interpolation. This is due to the fact that the sampling grid of a chrominance component has a lower resolution than that of the luminance component.

The H.264 also has other interesting features in computing motion vectors.

These vectors may be drawn from points outside the image area. In this case, the reference frame is required to be extrapolated beyond its boundaries. Motion vectors may also be estimated from multiple reference images coded previously.

In addition to the motion-compensated MB modes described earlier, a P macroblock may also be coded in the *skip mode* ( *P_Skip* type). In this mode the MB is simply copied from its previously coded frame.

### 1.5.3.6 Inter-frame Prediction in B Slices

In H.264 the concept of B slices is further generalized in comparison with the usual compression algorithm. In this case, B slices use two distinct lists of reference pictures. They are referred to as the first $list_0$ and the second $list_1$ reference picture lists, respectively. In B slices, there are four additional types of inter-picture prediction, namely, $list_0$, $list_1$, bi-predictive, and direct prediction, besides the usual prediction modes of *P-Inter Frame Slices*. In this case also an MB may be skipped if the prediction error remains the same as the corresponding MB of the previous encoded frame.

### 1.5.3.7 Integer Transform and Scaling

The H.264 compression technique is distinguished by the use of *integer cosine transforms* (ICT) [18] and, that too, for $4 \times 4$ blocks (instead of the usual $8 \times 8$ DCTs). The ICT has the following advantages compared to the DCT:

1. It is an integer transform. All operations can be carried out with 16-bit integer arithmetic without loss of accuracy.

2. The inverse transform is fully specified in the H.264 standard and, if this specification is followed correctly, mismatch between encoders and decoders should not occur.

3. The core part of the transform is free of multiplications, that is, it only requires additions and shifts.

4. A scaling multiplication (part of the complete transform) is integrated into the process of quantization (reducing the total number of multiplications).

Moreover, the ICT can easily be related to the DCT. The $4 \times 4$ DCT of an input array X is given in Eq. (1.16).

$$
Y = AXA^T = \begin{bmatrix} a & a & a & a \\ b & c & -c & -b \\ a & -a & -a & a \\ c & -b & b & -c \end{bmatrix} [X] \begin{bmatrix} a & b & a & c \\ a & c & -a & -b \\ a & -c & -a & b \\ a & -b & a & -c \end{bmatrix}, \quad (1.16)
$$

where $a = \frac{1}{2}, b = \sqrt{\frac{1}{2}}cos(\frac{\pi}{8}), c = \sqrt{\frac{1}{2}}cos(\frac{3\pi}{8})$.

We may modify the matrix $A$ in such a way that it consists of integer values only, though its orthogonal property remains intact. This is what is done with the matrix $T_4$ in Eq. (1.17). In this case, the 2nd and 4th rows of matrix $T_4$, and the 2nd and 4th columns of matrix $T_4^T$ are scaled by a factor of 2, and the postscaling matrix E is scaled down for compensation. E is a matrix of scaling factors, and the symbol '$\otimes$' indicates elementwise multiplication. The final forward transform is given in Eq. ( 1.17).

$$
\begin{aligned}
Y &= (T_4 X T_4^T) \otimes E \\
&= \left( \begin{bmatrix} 1 & 1 & 1 & 1 \\ 2 & 1 & -1 & -2 \\ 1 & -1 & -1 & 1 \\ 1 & -2 & 2 & -1 \end{bmatrix} [X] \begin{bmatrix} 1 & 2 & 1 & 1 \\ 1 & 1 & -1 & -2 \\ 1 & -1 & -1 & 2 \\ 1 & -2 & 1 & -1 \end{bmatrix} \right) \\
&\otimes \begin{bmatrix} a^2 & \frac{ab}{2} & a^2 & \frac{ab}{2} \\ \frac{ab}{2} & \frac{b^2}{4} & \frac{ab}{2} & \frac{b^2}{4} \\ a^2 & \frac{ab}{2} & a^2 & \frac{ab}{2} \\ \frac{ab}{2} & \frac{b^2}{4} & \frac{ab}{2} & \frac{b^2}{4} \end{bmatrix} ,
\end{aligned}
\tag{1.17}
$$

where $a = \frac{1}{2}, b = \sqrt{\frac{2}{5}}$. Elements of $E$ are known as *postscaling factors*, which can be any of the values given by $a^2$, $\frac{b^2}{4}$ ,or $\frac{ab}{2}$.

This transform is an approximation to the $4 \times 4$ DCT and is not identical to it. The inverse transform is given by Eq. (1.18).

$$
\begin{aligned}
X &= \begin{bmatrix} 1 & 1 & 1 & \frac{1}{2} \\ 1 & \frac{1}{2} & -1 & -1 \\ 1 & -\frac{1}{2} & -1 & 1 \\ 1 & -1 & 1 & -\frac{1}{2} \end{bmatrix} \left( [Y] \otimes \begin{bmatrix} a^2 & ab & a^2 & ab \\ ab & b^2 & ab & b^2 \\ a^2 & ab & a^2 & ab \\ ab & b^2 & ab & b^2 \end{bmatrix} \right) \\
&\begin{bmatrix} 1 & 1 & 1 & 1 \\ 1 & \frac{1}{2} & -\frac{1}{2} & -1 \\ 1 & -1 & -1 & 1 \\ \frac{1}{2} & -1 & 1 & -\frac{1}{2} \end{bmatrix} .
\end{aligned}
\tag{1.18}
$$

### 1.5.3.8  Quantization

The H.264 uses scalar quantization as given by

$$
Z(i, j) = round \left( W(i, j) \frac{E(i, j)}{Qstep} \right),
\tag{1.19}
$$

where $W(i, j)$ is the $(i, j)$th coefficient of the $4 \times 4$ ICT of $X$ and it is computed as $W = T_4 X T_4^T$. $E(i, j)$ is the corresponding element of the post-scaling matrix. $Qstep$ is the quantization step size, and $Z(i, j)$ is the corresponding quantized coefficient. The standard supports a total of 52 values for the parameter $Qstep$.

### 1.5.3.9    Second Transformation of DC Coefficients

As $Intra\_16 \times 16$ prediction modes and chrominance intra modes are intended for coding smooth areas, the DC coefficients of each $4 \times 4$ residual error blocks (as obtained by the first stage of transformation through prediction from neighboring blocks) undergo a second transform. In a $16 \times 16$ MB, the DC coefficient of each $4 \times 4$ block is transformed using a $4 \times 4$ Hadamard transform, and the DC coefficients of each $4 \times 4$ block of chrominance coefficients are grouped in a $2 \times 2$ block and are further transformed in the same way prior to their quantization. The transformations of corresponding DC luminance values and chrominance values are shown below in Eqs. (1.20) and (1.21), respectively.

$$Y_D = \frac{1}{2} \left( \begin{bmatrix} 1 & 1 & 1 & 1 \\ 1 & 1 & -1 & -1 \\ 1 & -1 & -1 & 1 \\ 1 & -1 & 1 & -1 \end{bmatrix} [W_D] \begin{bmatrix} 1 & 1 & 1 & 1 \\ 1 & 1 & -1 & -1 \\ 1 & -1 & -1 & 1 \\ 1 & -1 & 1 & -1 \end{bmatrix} \right). \quad (1.20)$$

$$Y_D = \frac{1}{2} \left( \begin{bmatrix} 1 & 1 \\ 1 & -1 \end{bmatrix} [W_D] \begin{bmatrix} 1 & 1 \\ 1 & -1 \end{bmatrix} \right). \quad (1.21)$$

In the above equations, $W_D$ is the block of DC coefficients, and $Y_D$ is the block after transformation, which is subjected to quantization subsequently.

### 1.5.3.10    Entropy Coding

In H.264, syntax elements and quantized transform coefficients are separately entropy coded following two different *variable length code* (VLC) tables. The syntaxes used a single adaptive code table that is customized according to the data statistics. It is essentially an exp-Golomb code [155] with very simple and regular decoding properties. There is also provision to make these tables dynamic depending on the change in data statistics. For quantized transform coefficients, a more efficient method called *context-adaptive variable length coding* (CAVLC) [155] is employed.

### 1.5.3.11    In-Loop Deblocking Filter

In-loop deblocking filter is used to reduce the blocking artifacts that may appear due to the block-based encoding of data. Usually, block edges are poorly reconstructed with this type of encoding. That is why H.264 defines an adaptive in-loop deblocking filter where the strength of filtering is controlled by the values of several syntax elements.

### 1.5.3.12    Network Abstraction Layer

H.264 also defines how data should be formatted for its interface with different transport layers in its specification for NAL. The NAL units define the partitioning and packaging of this data stream. Each of this unit contains an

(a)                                                    (b)

**Figure 1.27**: Image resizing from (a) HDTV (of frame-size $1080 \times 1920$) to (b) NTSC (of frame-size $480 \times 640$).

integer number of bytes and is employed in specifying a generic format for use in both packet-oriented and bitstream systems.

## 1.6  Examples of a Few Operations in the Compressed Domain

A number of direct implementation of various image and video processing operations in the compressed domain have been advanced. As this book mainly deals with the fundamentals and issues related to the development of those techniques, we provide an overview of such operations.

1. **Image Resizing:** In different applications, before the display, storage, or transmission of images, resizing images may be needed to fit them according to the resolutions of display devices or a format of storage or the limited duration of transmission. For example, we may need to convert an image in NTSC format into an HDTV one and vice versa. In Internet applications it is quite common to transmit an image with varying resolutions and sizes in keeping with different specifications of the display and communication network at client ends. There are quite a few techniques reported in the literature for performing this resizing operation directly in the compressed domain, in particular for images represented in the block DCT domain (as wavelet representation has the advantage of multiresolution representation of images, and resizing operations become trivial with that representation). A typical example of conversion of an image in HDTV format into an NTSC one in the block DCT space [103] is shown in Figure 1.27.

2. **Transcoding:** Transcoding is an important task that is often required for providing multimedia interoperability and bringing heterogeneous

(a)                                          (b)

**Figure 1.28**: Color image enhancement in the block DCT space: (a) Original and (b) Enhanced image.

devices together. In fact, resizing is also a type of transcoding operation, in which images in one spatial resolution are converted into another as demonstrated in Figure 1.27. On the other hand, transcoding of an image or video from one compression standard into another is often required. For example, we may consider transcoding JPEG2000 images into JPEG format or vice versa. Similarly, to provide higher interoperability in a video player, we may consider transcoding of H.264 video into MPEG-2. The brute force transcoding would have used the decoder and encoder in succession for converting image (or video) from one standard into the other. However, as discussed previously, compressed domain operations have several advantages, and various techniques [144] have been put forward by researchers to perform transcoding in the compressed domain itself.

3. **Image Enhancement:** Image enhancement is a task often aimed at rendering images with better visual perceptibility. Due to various reasons, raw image data require processing before display. For example, it may have a small dynamic range of intensity values on account of the presence of strong background illumination or due to insufficient lighting. A large dynamic range may also create problems caused by the limitation of the displayable range of colors or brightness values of a display device. In fact, it is more challenging in the event of imaging under a widely varying scene illumination. Image enhancement attempts to take care of these situations mainly by adjusting contrast, brightness and color of images. Though there are various methods reported in the spatial domain, it has been observed that compressed domain operations are also capable of performing these tasks usually with less computation but with comparable performance. A typical example of color image enhancement in the block DCT domain [105] [1] is shown in Figure 1.28 (**see color insert**).

---

[1]The     original     image     was     obtained     from     the     web-site *http://dragon.larc.nasa.govt/retinex/pao/news.*

4. **Information Hiding:** Due to the revolution in information and communication technology, there is an explosion of information generation and exchange. In particular, production and management of multimedia documents have become an integral part of today's social interaction. Under its aegis, covert communication of messages through these different digital media such as audio, text, images, videos, etc., has also become an issue not only from a social consideration but also from a technological point of view. This has led to the development and analysis of various techniques of information hiding. There are two different types of tasks involved in this domain. The first is *digital watermarking* [30], mainly used for copyright protection and authentication of digital media, and the other is *steganography* [30, 70], the art of covert communication with usual means (of unsuspected documents). For images and videos, as mainly these documents are available in the compressed domain, there is a considerable thrust toward developing algorithms of those two major operations and their analysis in the compressed domain itself.

5. **Image and Video Analysis:** Besides the foregoing three examples, various efforts are being made to develop different image and video analysis operations directly in the compressed domain. Some of them include video and image editing operations, image registration, object matching, video indexing, etc.

## 1.7    Issues and Performance Measures

While developing algorithms in the compressed domain, we should always consider the cost-benefit analysis of the technique compared to techniques realizing similar objectives in the spatial domain. There are three major issues in this regard as discussed in the following text.

### 1.7.1    Complexity of Algorithms

While expressing the complexities of algorithms, we have to consider both savings in computation and storage cost compared to equivalent computations in the spatial domain. In many cases, while expressing computational complexity, instead of determining the order of the complexity functions of input data size, the absolute number of operations per unit of input or output data (pixel for images) is used for measuring computational complexity. This is due to the fact that, in many cases, the order of complexity remains the same, though in absolute terms there are gains in computation. For example, while multiplying two matrices, we may reduce the number of multiplications if some of the elements of these matrices are zero and some of them are repeated. Exploiting this sparseness in a matrix, we may design an efficient computation for certain operations. Using such properties of matrices we may design a specific computational technique for performing operations. As matrix multiplication is quite a common operation in compressed domain processing, we present here a *general computational complexity model* [103] for finding the number of multiplications and additions, assuming that some of the elements are zeroes and repeated in a row.

Let $A$ be a sparse matrix of size $L \times N$, and it is to be multiplied with another arbitrary matrix $B$ of size $N \times M$. If $z_i$ is the number of *zero* elements, and $d_i$ is the number of *distinct* elements (of distinct magnitudes only) in the $i$th row of $A$, the total number of multiplications ($n_m(.)$) and the total number of additions ($n_a(.)$) are given by following equations:

$$n_m(A; L, N, M) = M\Sigma_{i=1}^{L} d_i. \tag{1.22}$$

$$n_m(A; L, N, M) = M\Sigma_{i=1}^{L}(N - z_i - 1). \tag{1.23}$$

Let us refer to this model as the *general computational complexity model of matrix multiplication (GCCMMM)* for computing the number of multiplications and additions in multiplying two sparse matrices. In this model, however, elements of $B$ are ignored, as $B$ is considered here as the variable (say, an input block), while $A$ is a fixed matrix operator. By exploiting other properties and constraints, further reductions in these numbers are possible. For example, if the value of an element is 1, we can ignore multiplication with that number. This is also true for an element expressed in integral powers of 2 (for both positive and negative numbers in the exponent). In that case, the multiplication (or division) operation is carried out by left (or right) shift of the binary representation of the number. However, the *GCCMMM* provides the upper estimate of the computational requirement for matrix multiplication.

### 1.7.2    Quality of Processed Images or Videos

The performance of different algorithms should also be judged by considering the quality of the processed images or videos. There are two distinct approaches while judging the quality of processed images. The first one attempts to find out the similarity between a reference image (the benchmark or ideal one) and the processed one. In the second approach, quality is judged by considering the visibility of different artifacts such as blurring, blocking, and ringing artifacts, which may occur due to processing. Though in the first case one requires a reference image for computing the metrics, in the second approach the measure may be defined solely on the intrinsic characteristics of images or videos. The latter type of measures or metrics is known as *no-reference metrics*, whereas metrics based on reference data are known as *full reference metrics*. We review next some of the measures used for judging the quality of images.

#### 1.7.2.1  Similarity with respect to a Benchmark or Reference Image

Several measures have been proposed for expressing the similarity between two images. Some of them are discussed below.

1. **Peak Signal-to-Noise Ratio (PSNR):** Given a benchmark image, we may consider the faithful reconstruction of the image as a measure of quality. In this regard, for images (and video frames), PSNR is often used as a simple fidelity metric. It is defined below for a color image.

   Let $I_s(x, y), s = 1, 2, 3$ be the color components of the benchmark images of size $M \times N$, and $I_s'(x, y), s = 1, 2, 3$ be the respective reconstructed color components. Then the PSNR is defined as

   $$PSNR = 20log \left( \frac{255}{\sqrt{\frac{\Sigma_{\forall s} \Sigma_{\forall x} \Sigma_{\forall y} (I_s(x,y) - I_s'(x,y))^2}{3MN}}} \right) \qquad (1.24)$$

   In the above equation, it is assumed that the peak signal value (or the maximum brightness value in the image) is 255. Accordingly, the PSNR for gray-level or monochrome images is also defined.

2. **Universal Quality Index (UQI):** We may easily observe that the PSNR is not necessarily a good measure of the similarity of images. For example, given a shift in the image pixels, the resulting image will be having a very poor PSNR value with respect to its original one, even though both of them appear to be very similar. We denote this type of similarity as perceptual similarity. There are different metrics proposed in the literature for measuring perceptual similarity or the quality of a processed image. One such measure is the *universal quality index* [170] defined next.

   Let $x$ and $y$ be two distributions. The UQI between these two distributions is defined as:

   $$UQI(x, y) = \frac{4\sigma_{xy}^2 \bar{x}\bar{y}}{(\sigma_x^2 + \sigma_y^2)(\bar{x}^2 + \bar{y}^2)}, \qquad (1.25)$$

   where $\sigma_{xy}^2$ is the covariance between $x$ and $y$, $\sigma_x$ and $\sigma_y$ are the standard deviations of $x$ and $y$, respectively. $\bar{x}$ and $\bar{y}$ are their respective means. It may be noted that this measure takes into account the correlation between the two distributions and also their proximity in terms of brightness and contrast. The UQI values should lie in the interval $[-1, 1]$. Processed images with UQI values closer to 1 are more similar in quality according to our visual perception. Usually, the UQI is computed locally for two images, and then the average over the set of local observations is considered the overall UQI between two images.

3. **Structural Similarity(SSIM):** The UQI has been further generalized in the following form [152] and named *Structural Similarity* measure.

$$SSIM(x,y) = \frac{(2.\sigma_{xy}^2 + C_2)(2.\bar{x}\bar{y} + C_1)}{(\sigma_x^2 + \sigma_y^2 + C_2)(\bar{x}^2 + \bar{y}^2 + C_1)}, \qquad (1.26)$$

where all the previous notations of Eq. (1.25) are defined in the same way. $C_1$ and $C_2$ are new parameters, which are functions of the dynamic range of the image (say $L$) and expressed as $(K_1L)^2$ and $(K_2L)^2$, and $(K_1, K_2 \ll 1)$, respectively. In [152], $K_1$ and $K_2$ are taken as 0.01 and 0.03, respectively.

It can be observed that, for $C_1 = C_2 = 0$, the SSIM becomes the same as the UQI. One advantage of the SSIM is that it provides more stability in numerical computation by introducing small positive biases in the factors of the numerator. Like UQI, the SSIM is also preferred for computing local similarities and then averaged over the set of observations.

### 1.7.2.2   Visibility of Artifacts

Quality is also judged by the appearance or suppression of different types of artifacts in the processed image while comparing it with the input or reference image. Even without any comparison, an image itself can be attributed by the existence and strength of such artifacts. There are different types of artifacts that are usually observed in images, namely, *blurring*, *blocking*, and *ringing artifacts*. *Blurring* of edges in an image occurs due to the suppression of its high spatial frequency components, while *ringing* around an edge takes place due to the quantization of high-frequency coefficients in transform coding. In a blurred edge, it gets smeared and loses sharpness, whereas *ringing* is visible as *ripples around sharp edges* [96]. *Blocking artifacts* are visible in a block-based compression scheme like JPEG. It occurs due to discontinuity in the block boundaries introduced by independent processing of blocks such as quantization of transform coefficients. There are different models proposed for measuring these artifacts. Two such typical techniques are discussed here.

1. **Blur and Ringing Metrics:** In [96], the blur of a vertical edge is measured by its width, which is computed as the distance between the local maximum and the local minimum in the neighborhood of the edge point. The average width of vertical edge points provides the measure of the blur of an image. The measure could have been defined including edges in other directions (e.g., horizontal edges, etc.). However, it is observed that there is only a small difference in expressing the relative perceptual

quality of images in both cases. For the convenience of computation, vertical edges are considered. The blur metric could be both a noreference or a full-reference metric. In the latter case, edge points are determined from the reference image, whereas for the noreference metric they are determined from the image (or processed image) itself.

In [96] *ringing* is measured by using vertical edges only. In this case, first an interval supporting the ringing phenomena is defined around an edge point. The support to the left (right) of the edge point is defined by the set of points in the horizontal line within a fixed width (or distance) excluding the blurring points (points within *blurring width* as defined in expressing its blur previously). Then the left (right) ringing is defined as the length of the left (right) support multiplied by the dynamic range of the difference signal between the processed and original one within the support. The ringing at an edge point is defined as the sum of *left* and *right ringing*, and for the whole image it is the average of ringing of vertical edge points. *Ringing metric* (RM) is a full-reference metric. It may be noted that both blurring and ringing measures are typically used for judging the quality of JPEG2000 compressed images.

2. **JPEG Quality Metric (JPQM):** Wang et al. [153] proposed an interesting *no-reference metric* for judging the image quality reconstructed from the block DCT space to take into account visible blocking and blurring artifacts. This metric computes (i) average horizontal and vertical discontinuities at corresponding $8 \times 8$ block boundaries, (ii) activity in images expressed by deviations of average horizontal and vertical gradients from their respective average block discontinuities, and (iii) the average number of zero-crossings in these gradient spaces. Subsequently, these factors are combined in a nonlinear expression to produce the resulting value of the metric. It may be noted, that for an image with good visual quality, the JPQM value is close to 10.

### 1.7.2.3 Measure of Colorfulness

Colorfulness is also another attribute for judging the quality of a colored image. In [139], Susstrunk and Winkler proposed a no-reference metric called *colorfulness metric* (CM) for measuring the diversity and contrast of colors in an image. The definition for this metric in the RGB color space is as given below.

Let the red, green, and blue components of an Image $I$ be denoted as $R$, $G$, and $B$, respectively. Let $\alpha = R - G$, and $\beta = (\frac{R+G}{2}) - B$. Then the colorfulness of the image is defined as

$$CM(I) = \sqrt{\sigma_\alpha^2 + \sigma_\beta^2} + 0.3\sqrt{\mu_\alpha^2 + \mu_\beta^2}, \tag{1.27}$$

where $\sigma_\alpha$ and $\sigma_\beta$ are standard deviations of $\alpha$ and $\beta$, respectively. Similarly, $\mu_\alpha$ and $\mu_\beta$ are their means.

In [105], this metric has been converted to a *full-reference* one by comparing the color with respect to the original image, and it is expressed as the ratio of CMs between the processed image and its original. The new measure is named the *color enhancement factor* (CEF).

### 1.7.3 Level of Compression of the Input and Output Data

While judging the quality of the processed images due to compressed domain processing, we observe degrees of compression for both the input and output of a processing. The performance of a scheme may vary with the level of compression of images and videos. The level of compression is usually measured by the *compression ratio*, which is expressed as the ratio between the size of raw input data and the size of compressed data. The compression ratio may also be expressed in percentage (%). Further, assuming 8 bits per pixel for a gray-level image and 24 bits per pixel for color images, compression levels are expressed by the average *bits per pixel (bpp)* required for representing the compressed data. For video, the preferred unit in this case is the *bits per second (bps)*, by taking care of the *frame rate* of a video.

---

## 1.8 Summary

In the compressed domain, images and videos are represented in alternative forms that are quite different from their representations in the spatial or spatiotemporal domain. In today's world, due to the digital revolution in our daily life, in particular due to the stupendous advancement in the multimedia and communication technology, these images and videos are increasingly available in the compressed format. Compression standards such as JPEG, JPEG2000, MPEG-2, MPEG-4, H.264, etc., are not known only to specialists but they have gained popularity among modern consumers and users as well. Hence, in applications involving image and video processing, computations directly in the domain of alternative representation are often felt necessary as it reduces the time and storage required for the processing. In this chapter, with an introduction to different popular compression technology, motivation and performance issues related to the compressed domain processing have been discussed.

# Chapter 2

## Image Transforms

Image transforms [52] are used for decomposing images in different structures or components so that a linear combination of these components provide the image itself. We may consider an image as a 2-D *real* function $f(x, y) \in L^2(\mathbb{R}^2)$, $(x, y) \in \mathbb{R}^2$, where $\mathbb{R}$ is the set of real numbers and $L^2(\mathbb{R}^2)$ is the space of all square integrable functions. In a *transform*, the function is represented as a linear combination of a family (or a set) of functions, known as *basis functions*. The number of basis functions in the set may be *finite* or *infinite*. Let $B = \{b_i(x, y)| -\infty < i < \infty\}$ be a set of basis functions, where $b_i(x, y)$ may be in *real* ($\mathbb{R}$) or *complex space* ($\mathbb{C}$). In that case, a transform of the function $f(x, y)$ with respect to the basis set $B$ is defined by the following expression:

$$f(x, y) = \sum_{i=-\infty}^{\infty} a_i b_i(x, y), \qquad (2.1)$$

where $a_i$'s are in $\mathbb{R}$ or $\mathbb{C}$ depending upon the space of the basis functions. They are known as *transform coefficients*. This set of coefficients provides an alternative description of the function $f(x, y)$.

In this chapter we discuss a number of different image transforms and their properties. As all our transforms are extended from their counterparts in 1-D, for the sake of simplicity, we initially restrict our discussion to 1-D. Their extensions in 2-D are discussed subsequently.

## 2.1   Orthogonal Expansion of a Function

Let $f(x) \in L^2(\mathbb{R})$ be a function over the support $[a, b] \subseteq \mathbb{R}$, and $B = \{b_i(x)| -\infty < i < \infty, x \in [a, b]\}$ be a set of basis functions (either in $\mathbb{R}$ or $\mathbb{C}$). In our discussion we consider every function to be *integrable* in $[a, b]$. An expansion of $f(x)$ with respect to $B$ is defined as follows:

$$f(x) = \sum_{i=-\infty}^{\infty} \lambda_i b_i(x), \tag{2.2}$$

where $\lambda_i$'s are in $\mathbb{R}$ or $\mathbb{C}$ depending upon the space of the basis functions.

**Definition 2.1 Inner Product:** *The inner product of two functions $h(x)$ and $g(x)$ in $[a, b] \subseteq \mathbb{R}$ is defined as*

$$< h, g >= \int_a^b h(x)g^*(x)dx, \tag{2.3}$$

*where $g^*(x)$ is the complex conjugate of $g(x)$.* □

**Definition 2.2 Orthogonal Expansion:** *An expansion of a function $f(x)$ in the form of Eq. (2.2) is orthogonal when the inner products of pairs of basis functions in $B$ have the following properties:*

$$\begin{aligned} < b_i, b_j > &= 0, \quad \text{for } i \neq j, \\ &= c_i, \quad \text{Otherwise (for } i = j), \text{ where } c_i > 0. \end{aligned} \tag{2.4}$$

□

**Theorem 2.1** *Given an orthogonal expansion of $f(x)$ with a base $B$, transform coefficients are computed as follows:*

$$\lambda_i = \frac{1}{c_i} < f, b_i > . \tag{2.5}$$

**Proof:**

$$\begin{aligned} < f, b_i > &= \int_a^b f(x)b_i^*(x)dx &&\text{(from Def. 2.1)} \\ &= \int_a^b \left( \sum_{j=-\infty}^{\infty} \lambda_j b_j(x) \right) b_i^*(x)dx &&\text{(from Eq. (2.2))} \\ &= \lambda_i \int_a^b b_i(x)b_i^*(x)dx + 0 \\ &= \lambda_i c_i \end{aligned} \tag{2.6}$$

□

**Corollary 2.1** *For an orthogonal base, if $c_i = 1$ for all the $i$-th basis function, we can simply obtain the transform coefficients as*

$$\lambda_i =< f, b_i > . \tag{2.7}$$

□

Such a family of basis functions is known as *orthonormal*, and corresponding basis functions are called *orthonormal basis functions*. Any subset of orthogonal base $B$ also satisfies the property of orthogonality as expressed in Eq. (2.4). However, it may not be sufficient or *complete* for full reconstruction of the function $f(x)$. The set of orthogonal (orthonormal) functions that is complete for every $f(x)$ is known as *complete orthogonal (orthonormal) base*.

The set of transform coefficients $\Lambda = \{\lambda_i| - \infty < i < \infty\}$ provides the alternative description of $f(x)$ and is known as its *transform*, while the *space* of these coefficients is termed as the *transform space*. Eq. (2.5) or (2.7) denotes the *forward transform*, while the expansion with transform coefficients (Eq. (2.2)) is termed as the *inverse transform*. In continuous space for $i$, Eq. (2.2) is expressed as

$$f(x) = \int_{i=-\infty}^{\infty} \lambda_i b_i(x) di. \tag{2.8}$$

One of the important facts about the *orthonormal expansion* of the function is that it preserves the energy of the function as expressed in the following theorem.

**Theorem 2.2** *For an orthonormal expansion of the function as given in Eq. (2.2), the sum of squares of the transform coefficients remains the same as the sum of squares of functional values within its support (or energy of the function) as expressed in the following relation*

$$\sum_{i=-\infty}^{\infty} \lambda_i^2 = \int_a^b f^2(x) dx. \tag{2.9}$$

**Proof:**

$$
\begin{aligned}
\int_a^b f^2(x) dx &= \int_a^b f(x) f^*(x) dx, \\
&= \int_a^b f(x) \left( \sum_{i=-\infty}^{\infty} \lambda_i^* b_i^*(x) \right) dx, \\
&= \sum_{i=-\infty}^{\infty} \lambda_i^* \int_a^b f(x) b_i^*(x) dx, \\
&= \sum_{i=-\infty}^{\infty} \lambda_i^* \lambda_i, \\
&= \sum_{i=-\infty}^{\infty} \lambda_i^2.
\end{aligned}
\tag{2.10}
$$

$\square$

In the following text we examine a few examples of *orthogonal transforms*.

## 2.1.1 Trivial Expansion with Dirac Delta Functions

**Definition 2.3** *The **dirac delta function** $\delta(x)$ has a unit infinite impulse at $x = 0$ and zero elsewhere. It holds the following properties:*

$$
\begin{aligned}
\int_{-\infty}^{\infty} \delta(x)dx &= 1, \\
\int_{-\infty}^{\infty} f(x)\delta(x - x_0)dx &= f(x_0).
\end{aligned}
\tag{2.11}
$$

□

Hence the function $f(x)$ in the support of $\mathbb{R}$ can be written as

$$
f(x) = \sum_{i=-\infty}^{\infty} f(i)\delta(x - i).
\tag{2.12}
$$

## 2.1.2 Fourier Series Expansion

A periodic function $f(x)$ with a period of length $\tau$ can be expanded using *Fourier series expansion* [93]. In this case, it is sufficient to define $f(x)$ over a period, say, in $[-\frac{\tau}{2}, \frac{\tau}{2}]$. The orthogonal base in this support consists of the base $\{e^{j2\pi\frac{m}{\tau}x}|m \in \mathbb{Z}\}$, where $j$ is the complex number $\sqrt{-1}$ and $\mathbb{Z}$ is the set of integers. We may check the orthogonality of this base by the following:

$$
\int_{-\frac{\tau}{2}}^{\frac{\tau}{2}} e^{j2\pi\frac{m}{\tau}x}dx = \left[\frac{e^{j2\pi\frac{m}{\tau}x}}{j2\pi\frac{m}{\tau}}\right]_{-\frac{\tau}{2}}^{\frac{\tau}{2}} = \begin{cases} \tau, & \text{for } m = 0, \\ 0, & \text{Otherwise.} \end{cases}
\tag{2.13}
$$

Hence, the Fourier series of $f(x)$ is given by

$$
\lambda_m = \frac{1}{\tau}\int_{-\frac{\tau}{2}}^{\frac{\tau}{2}} f(x)e^{-j2\pi\frac{m}{\tau}x}dx.
\tag{2.14}
$$

In this case, $\lambda_m$'s are complex numbers. In fact as $e^{j2\pi\frac{m}{\tau}x} = cos(2\pi\frac{m}{\tau}x) + jsin(2\pi\frac{m}{\tau}x)$, *Fourier series* can also be expanded by *sine* and *cosine harmonics*, forming another form of the orthogonal base in this case.

If $f(x)$ is an *even function*, only *cosine harmonics* ($\{cos(2\pi\frac{m}{\tau}x)|m \in \mathbb{N}\}$) are sufficient to form the *base* of its expansion ($\mathbb{N}$ being the set of nonnegative integers). On the other hand, for an *odd function* $f(x)$, it is the set of *sine harmonics* ($\{sin(2\pi\frac{m}{\tau}x)|m \in \mathbb{N}, m \neq 0\}$) that can construct it.

## 2.1.3 Fourier Transform

As $\tau$ tends to $\infty$ in Eq. (2.14), the transform is applicable for any function with its support being the complete *real* axis or $[-\infty, \infty]$. Let us rename $2\pi\frac{m}{\tau}$

as $\omega$, a radian frequency variable. As $\tau \to \infty$, $\frac{1}{\tau}(= \frac{\delta\omega}{2\pi}) \to 0$. In that case, the Fourier series is obtained as

$$\lambda(\omega) = \frac{\delta\omega}{2\pi} \int_{-\infty}^{\infty} f(x)e^{-j\omega x}dx. \qquad (2.15)$$

We may express the corresponding factor in the expansion for a continuously varying $\omega$ in the interval of $[\omega_0 - \frac{\delta\omega}{2}, \omega_0 + \frac{\delta\omega}{2}]$ as $\lambda(\omega_0) = \hat{f}(j\omega_0)\frac{\delta\omega}{2\pi}$. This leads to the definition of the *Fourier transform* of $f(x)$ as given below:

$$\mathbb{F}(f(x)) = \hat{f}(j\omega) = \int_{-\infty}^{\infty} f(x)e^{-j\omega x}dx. \qquad (2.16)$$

The *inverse Fourier transform* of $\hat{f}(j\omega)$ (provided it is *integrable*) is given by

$$f(x) = \frac{1}{2\pi} \int_{-\infty}^{\infty} \hat{f}(j\omega)e^{j\omega x}d\omega. \qquad (2.17)$$

In our notation, we consistently follow the convention of denoting the Fourier transform of a function $g(x)$ as either $\mathbb{F}(g(x))$ or $\hat{g}(j\omega)$. The orthogonality of the base $\{e^{j\omega x} | \omega \in \mathbb{R}\}$ over $\mathbb{R}$ is expressed in the following form:

$$\int_{-\infty}^{\infty} e^{j\omega x}dx = \begin{cases} 2\pi\delta(\omega), & \text{for } \omega = 0, \\ 0, & \text{otherwise.} \end{cases} \qquad (2.18)$$

### 2.1.3.1   Properties of Fourier Transform

The Fourier transform provides the amount of *oscillation* a function has at the *complex frequency* $j\omega$. The contribution of the complex sinusoid $e^{j\omega x}$ is expressed by $\hat{f}(j\omega)\frac{\delta\omega}{2\pi}$. Hence, $\hat{f}(j\omega)$ is a *density* function. As it is a *complex* number, it may also be represented by its *magnitude* ($|\hat{f}(j\omega)|$) and *phase* ($\theta(\omega)$) such that $\hat{f}(j\omega) = |\hat{f}(j\omega)|e^{j\theta(\omega)}$, where both $|\hat{f}(j\omega)|$ and $\theta(\omega)$ are real functions over $\omega$ in $\mathbb{R}$ . As $f(x)$ is a *real* function, $\hat{f}(-j\omega) = \hat{f}^*(j\omega)$. This converts the complex frequency components at $\omega$ in the following *real* form:

$$\begin{aligned} \hat{f}(j\omega)e^{j\omega x} + \hat{f}^*(j\omega)e^{-j\omega x} &= |\hat{f}(j\omega)| \left( e^{j(\omega x + \theta(\omega))} + e^{-j(\omega x + \theta(\omega))} \right), \\ &= 2.|\hat{f}(j\omega)| \left( cos(\omega x + \theta(\omega)) \right). \end{aligned}$$
$$(2.19)$$

The above property makes the *magnitude* spectrum ($|\hat{f}(j\omega)|$) an *even* function of $\omega$, while the *phase* spectrum ($\theta(\omega)$) is an *odd* function of $\omega$. The domain of $\omega$ is known as the *frequency domain*, which is also referred to in this case as the *transform domain*, and the domain of $x$ is the original domain of the function, which usually refers to *time* for time-varying signals, or *space* for a space-varying function. In our discussion, we correlate the interpretation in the time domain with that in 1-D space. Some of the important properties of the Fourier transform [52, 93, 100] are listed below.

1. **Linearity:** The Fourier transform of a linear combination of functions is the same as the same linear combination of their transforms, that is,

$$\mathbb{F}(ag(x) + bh(x)) = a\hat{g}(j\omega) + b\hat{h}(j\omega). \tag{2.20}$$

2. **Duality:** If the Fourier transform of $f(x)$ is $g(j\omega)$, the Fourier transform of $g^*(jx)$ (i.e., complex conjugate of $g(jx)$) is $2\pi f(\omega)$.

3. **Reversal in space:** Reversal in space (or time) also causes reversal in the frequency domain. This implies that $\mathbb{F}(g(-x)) = \hat{g}(-j\omega)$.

4. **Translation:** Proportional shifts in phase occur across frequencies due to the translation in space (or time).

$$\mathbb{F}(g(x - X_0)) = e^{-j\omega X_0} \hat{g}(j\omega). \tag{2.21}$$

5. **Scaling:** Contraction in space (or time) by a factor $s$ causes scaling of frequencies by the same factor.

$$\mathbb{F}(g(\frac{x}{s})) = |s|\hat{g}(js\omega). \tag{2.22}$$

6. **Modulation:** The transform gets translated in the frequency space if the function is modulated (or multiplied) by a complex sinusoid.

$$\mathbb{F}(e^{j\omega_0 x} g(x)) = \hat{g}(j(\omega - \omega_0)). \tag{2.23}$$

As $cos(\omega_0 x) = \frac{1}{2}(e^{j\omega_0 x} + e^{-j\omega_0 x})$, the above relation can be expressed with the modulation of cosine function in the following way:

$$\mathbb{F}(g(x)cos(\omega_0 x)) = \frac{1}{2} \left( \hat{g}(j(\omega + \omega_0)) + \hat{g}(j(\omega - \omega_0)) \right). \tag{2.24}$$

7. **Convolution–multiplication theorem:** The *convolution* operation models the response of a *linear shift-invariant* (LSI) system. Given two functions $g(x)$ (an input) and $h(x)$ (the unit impulse response), the *convolution* between them is defined as

$$g \star h(x) = \int_{-\infty}^{\infty} g(\tau)h(x - \tau)d\tau. \tag{2.25}$$

The equivalence of a convolution operation in the *frequency* domain is simple multiplication of their transform, as shown below:

$$\mathbb{F}(g \star h(x)) = \hat{g}(j\omega)\hat{h}(j\omega). \tag{2.26}$$

8. **Multiplication-convolution Theorem:** Similar to the *convolution–multiplication theorem*, adhering to the *duality principle*, we may show that equivalent operation for *multiplication* of two functions *in space*

Table 2.1: Fourier transforms of a few useful functions

| Function | Description | Fourier transform |
|---|---|---|
| $\delta(x - X_0)$ | Translated impulse function | $e^{-j\omega X_0}$ |
| $e^{j\omega_0 x}$ | A single complex sinusoid | $2\pi\delta(\omega - \omega_0)$ |
| $\displaystyle\sum_{n=-\infty}^{\infty} \delta(x - nX_0)$ | Periodic impulses with a period of $x_0$ (*Dirac Comb*) | $\displaystyle\sum_{n=-\infty}^{\infty} e^{-jnX_0\omega} =$ $\displaystyle\frac{2\pi}{X_0} \sum_{k=-\infty}^{\infty} \delta(\omega - \frac{2\pi k}{X_0})$ (*Poisson Formula*) |
| $rect(x) = \begin{cases} 1 & \text{for } \|x\| < \frac{1}{2} \\ 0 & \text{Otherwise} \end{cases}$ | A rectangular box function | $sinc(\frac{\omega}{2}) = \frac{sin(\frac{\omega}{2})}{\frac{\omega}{2}}$ |
| $\frac{1}{2\pi} sinc(\frac{x}{2})$ | Sinc function as defined above | $rect(\omega)$ |
| $e^{-x^2}$ | Gaussian function | $\sqrt{\pi}e^{-\frac{\omega^2}{4}}$ |

(or *time*) is the *convolution* in the *frequency* domain. This relationship is expressed as follows:

$$\mathbb{F}(g(x)h(x)) = \frac{1}{2\pi}\hat{g}\star\hat{h}(j\omega). \tag{2.27}$$

9. **Differentiation in space or time domain:**

$$\mathbb{F}\left(\frac{d^n g(x)}{dx^n}\right) = (j\omega)^n \hat{g}(j\omega). \tag{2.28}$$

10. **Differentiation in frequency domain:**

$$\mathbb{F}^{-1}\left(j^n \frac{d^n \hat{g}(j\omega)}{d\omega^n}\right) = x^n g(x). \tag{2.29}$$

11. **Parseval's Theorem:**

$$\int_{-\infty}^{\infty} |g(x)|^2 dx = \frac{1}{2\pi} \int_{-\infty}^{\infty} |\hat{g}(j\omega)|^2 d\omega. \tag{2.30}$$

Fourier transforms of a few useful functions are listed in Table 2.1. It is interesting to note that the Fourier transform of a Gaussian function remains Gaussian in nature, and transform of a train of periodic impulses (Dirac deltas) of period $X_0$ in the spatial (temporal) domain corresponds to a similar form of periodic impulses in frequency domain with a period of $\frac{1}{X_0}$. The latter property is useful in proving the following theorem for the Fourier transform of a sampled function with a sampling period of $X_0$.

**Theorem 2.3** *Let $f(x)$ be sampled at a uniform sampling period of $X_0$ such that the resulting sampled function $f_s(x)$ is given as*

$$f_s(x) = \sum_{n=-\infty}^{\infty} f(nX_0)\delta(x - nX_0). \tag{2.31}$$

*Then the Fourier transform of $f_s(x)$ is as given below.*

$$
\begin{aligned}
\mathbb{F}(f_s(x)) &= \frac{1}{X_0} \sum_{k=-\infty}^{\infty} \hat{f}(j\omega)\delta\left(\omega - \frac{2\pi k}{X_0}\right), \\
&= \frac{1}{X_0} \sum_{k=-\infty}^{\infty} \hat{f}\left(j(\omega - \frac{2\pi k}{X_0})\right).
\end{aligned}
\tag{2.32}
$$

$\square$

The above theorem shows that, for a *lossless* recovery of a function from its uniformly sampled sequence (using an *ideal low-pass filter*), we need to sample it at a sampling frequency of at least *twice* the highest frequency component present in the function. For a *band-limited* signal, this should be at least twice its *bandwidth*. This is the *Nyquist rate* for sampling a signal or function without any error (known as *aliasing* error).

### 2.1.4 Shannon's Orthonormal Bases for Band-limited Functions

From the foregoing theorem, we find that the complete recovery of a signal $f(x)$ from the sequence of its sampled values at a uniform sampling interval of $X_0$, say, $f_s(x) = \{f(nX_0)| -\infty < n < \infty\}$, is possible if the sequence is filtered by an ideal low-pass filter whose Fourier transform $\hat{h}_{X_0}(\omega)$ is $rect(\frac{X_0\omega}{2\pi})$ (for definition of the function see Table 2.1). As the inverse transform of $\hat{h}_{X_0}(\omega)$ is $h_{X_0}(x) = \frac{sin(\frac{\pi x}{X_0})}{\frac{\pi x}{X_0}}$, $f(x) = f_s(x) \star h_{X_0}(x) = \sum_{n=-\infty}^{\infty} f(nX_0)h_{X_0}(x - nX_0)$. We should note that the family of functions $\{h_{X_0}(x - nX_0)| -\infty < n < \infty\}$ forms the orthonormal base (known as *Shannon's orthonormal base*) for the space of band-limited functions whose Fourier transforms are within the support of $[\frac{-\pi}{X_0}, \frac{\pi}{X_0}]$.

### 2.1.5 Wavelet Bases

Wavelets [93] are considered to be functions that should have ideally finite support in both its original domain and also in the transform domain (i.e., the frequency domain). However, in reality, no such function exists that has both these properties. For example, a *Dirac delta* $\delta(x)$ with an infinitesimally small support in space has the frequency response over the entire frequency range (in $[-\infty, \infty]$). On the other hand, a complex sinusoid such as $e^{j\omega_0 x}$

with an infinite support over $\mathbb{R}$ has the Fourier response of $\delta(\omega - \omega_0)$, which is again having the infinitesimally small support in the frequency domain. That there exists no such function is expressed by Heisenberg's uncertainty principle restated in the following form.

**Theorem 2.4** *For a function $f(x) \in L(\mathbb{R}^2)$, let the average location $\mu_x$ and average momentum $\mu_\omega$ be defined as follows:*

$$\begin{aligned} \mu_x &= \tfrac{1}{||f||^2} \int_{-\infty}^{\infty} x|f(x)|^2 dx, \\ \mu_\omega &= \tfrac{1}{2\pi||f||^2} \int_{-\infty}^{\infty} \omega|\hat{f}(\omega)|^2 d\omega. \end{aligned} \tag{2.33}$$

*Similarly, we measure the "uncertainty" in its "average deviations" in space (or time) and frequency by respective standard deviations $\sigma_x$ and $\sigma_\omega$ as defined below:*

$$\begin{aligned} \sigma_x^2 &= \tfrac{1}{||f||^2} \int_{-\infty}^{\infty} (x - \mu_x)^2|f(x)|^2 dx, \\ \sigma_\omega^2 &= \tfrac{1}{2\pi||f||^2} \int_{-\infty}^{\infty} (\omega - \mu_\omega)^2|\hat{f}(\omega)|^2 d\omega. \end{aligned} \tag{2.34}$$

*By drawing analogy from Heisenberg's uncertainty principle, we can prove the following [93]:*

$$\sigma_x^2 \sigma_\omega^2 \geq \frac{1}{4}. \tag{2.35}$$

$\square$

For ensuring good *localization* characteristics in both the original domain and transform domain, "wavelets" are aimed at keeping both the $\sigma_x$ and $\sigma_\omega$ low in the respective domains of functional representation. This provides a powerful tool for factorizing functions and analyzing them thereafter. From a function $\psi(x)$ that has these characteristics, we may generate a basis by considering the family of functions by translating and dilating $\psi(x)$ over the integral space $\mathbb{Z}$. This basis set may be represented as $\{\psi_{j,n}(x) = \frac{1}{\sqrt{2^j}} \psi(2^{-j}x - n)|j, n \in \mathbb{Z}\}$. We call $\psi(x)$ the *mother wavelet* for such a base. There exist mother wavelets that can generate *orthonormal bases* for representing a function not only in its original form but also with its approximations at *various lower resolutions*. The set of basis functions that are responsible for expanding the $j$th level of resolution is formed by translating the dilated wavelets $\psi_j(x)(= \frac{1}{\sqrt{2^j}} \psi(2^{-j}x))$ at discrete points of $\mathbb{Z}$. In this set, the mother wavelet itself is also scaled by the factor $2^{\frac{j}{2}}$ in its dilated form. In fact, the factors contributed by these basis functions at the $j$th resolution provide the variations in the function from its next higher-level resolution, that is, of $(j - 1)$th level of approximation of the function. For a better understanding of this decomposition of functions, we discuss the *multiresolution approximations* [92] of functions.

### 2.1.5.1   Multiresolution Approximations

In wavelet analysis of a function, the function is represented by its *approximation* and *variation* at a certain level of resolution. Here the term *resolution* refers to the *scale* of observation of functional values over a finite support. The higher the resolution, the lower is the scale. In the family of wavelet functions, at the $j$th level the scale of the function is considered as $2^j$, and hence the resolution may be treated as $2^{-j}$. The original representation of the function may be considered at level 0. To obtain an *approximation* of a function (say, $f(x)$) at level $j$, we may subject it to a low-pass filter that computes *an average at any point around its neighborhood of size proportional to $2^j$*. Let this approximation be denoted as $f_j(x)$. The *variations* are captured by *the difference between two such approximations at successive levels*, say, $d_j(x) = f_{j-1}(x) - f_j(x)$. An orthonormal wavelet expansion holds the following property.

$$d_j(x) = \sum_{n=-\infty}^{\infty} <f, \psi_{j,n}> \psi_{j,n}(x). \tag{2.36}$$

Like *variations* or *details*, approximations may also be represented by an orthonormal expansion of wavelet basis from a mother wavelet, say, $\phi(x)$, by dilating and translating it in the same way. Sometimes $\phi(x)$ is also referred to as the *scaling function*, and the approximation at the $j$th level may be obtained as follows:

$$f_j(x) = \sum_{n=-\infty}^{\infty} <f, \phi_{j,n}> \phi_{j,n}(x). \tag{2.37}$$

A few examples of such scaling functions are provided in the following text.

1. **Haar's wavelets**: This considers the wavelets as the piecewise constant functions over $[2^j n, 2^j (n+1)]$. In this case, the scaling function has a finite support; however its frequency response has infinite support, though it decays fast with increasing frequencies.

$$\phi(x) = \begin{cases} 1, & 0 \leq x \leq 1, \\ 0, & otherwise. \end{cases} \tag{2.38}$$

2. **Shannon's base**: Shannon's base has a support similar to Haar's in the frequency domain, which is a piecewise constant in the frequency range of $[2^{(j+1)}\pi n, 2^{(j+1)}\pi(n+1)]$.

$$\phi(x) = \frac{sin(\pi x)}{\pi x} \tag{2.39}$$

Corresponding to the Haar's scaling function, the mother wavelet for computing the details is

$$\psi(x) = \begin{cases} 1, & 0 \leq x < \frac{1}{2}, \\ -1, & \frac{1}{2} \leq x \leq 1, \\ 0, & otherwise. \end{cases} \tag{2.40}$$

### 2.1.5.2  Wavelet Bases for Multiresolution Approximations

These two families of basis functions, $\{\phi_j(x-n)|n \in \mathbb{Z}\}$ and $\{\psi_j(x-n)|n \in \mathbb{Z}\}$ form the complete orthonormal base for the function at the $(j-1)$th level of resolution. This signifies that the orthonormal base for a function $f(x)$ at its original resolution $(j = 0)$ is formed by the basis set $\{\frac{1}{\sqrt{2}}\phi(\frac{x}{2} - n), \frac{1}{\sqrt{2}}\psi(\frac{x}{2} - n)|n \in \mathbb{Z}\}$ given the two parent wavelets as $\phi(x)$ and $\psi(x)$. It should be noted that there exists a unique $\psi(x)$ given a $\phi(x)$ and vice versa. Their relationships are elaborated further during our discussion on *discrete wavelet transform* (DWT) (see Section 2.2.6).

## 2.2  Transforms of Discrete Functions

A *discrete function* is a sequence of functional values at *discrete points* regularly spaced in its domain. For example, the discrete representation of a function $f(x) \in L^2(\mathbb{R})$ could be in the form of $f(n) = \{f(nX_0)|n \in \mathbb{Z}\}$, where $X_0$ is the sampling interval and $f(n) \in L^2(\mathbb{Z})$. In any discrete representation, there is an implicit assumption of a uniform sampling interval or a sampling frequency. The *sampling theorem* provides the relationship between the discrete function and its continuous representation in the frequency domain. A discrete function $f(n)$ can also be considered as a *vector* in an *infinite dimensional vector space*. Discretization of a function may also be carried out over a finite interval with a finite number of samples in a sequence. For example, $\{f(n), n = 0, 1, 2, \dots, (N-1)\}$, represents a finite length sequence of $N$ samples. In this case, the sequence becomes a vector in an $N$-*dimensional space*. Let us represent this vector as $\vec{f} \in \mathbb{R}^N$. In our notation, sometimes we may also represent it as a column vector $\mathbf{f} = [f(0)f(1) \dots f(N-1)]^T$. $\mathbf{f}^T$ denotes the transpose of the matrix $\mathbf{f}$.

Any vector in the $N$-dimensional space could be represented by a linear combination of a set of $N$ *independent* vectors [45], say $B = \{\vec{b_i}|i = 0, 1, \dots, N-1\}$. Moreover, $B$ becomes an orthogonal set if the following property (as similar to the case of basis functions) holds true.

$$\vec{b_i^*}.\vec{b_j} \begin{array}{ll} = & 0, \quad \text{for } i \neq j, \\ = & c_i \quad \text{Otherwise (for } i = j), \text{ where } c_i > 0. \end{array} \quad (2.41)$$

In the above equation, the operator '.' denotes the *dot* product of two vectors. An *orthogonal basis set* $B$ is a *complete orthonormal set* if $||b_i|| = \vec{b_i^*}.\vec{b_i} = 1$ for every $b_i$.

In the expanded form, $\vec{f}$ is represented as

$$\vec{f} = \sum_{i=0}^{N-1} \lambda_i \vec{b_i}, \quad (2.42)$$

where $\lambda_i$'s are the magnitudes of the components along the direction $\overrightarrow{b_i}$ and computed as

$$\lambda_i = \overrightarrow{f}.\overrightarrow{b_i^*}. \tag{2.43}$$

Accordingly, the *transform* of **f** is defined by the vector $\mathbf{l} = [\lambda_0 \ \lambda_1 \ \ldots \ \lambda_{N-1}]^{\mathbf{T}}$ and can be expressed in the matrix form in the following way:

$$\mathbf{l} = \mathbf{Bf}, \tag{2.44}$$

where,

$$\mathbf{B} = \begin{bmatrix} \mathbf{b_0}^{*\mathbf{T}} \\ \mathbf{b_1}^{*\mathbf{T}} \\ \cdot \\ \cdot \\ \cdot \\ \mathbf{b_{N-1}}^{*\mathbf{T}} \end{bmatrix}. \tag{2.45}$$

In the above equation, **B** is the transform matrix, and $\mathbf{b_i}$ is the column vector corresponding to $\overrightarrow{b_i}$. If **B** is invertible, **f** could be recovered from **l** as

$$\mathbf{f} = \mathbf{B}^{-1}\mathbf{l}. \tag{2.46}$$

For an orthonormal transform matrix, **B**, $\mathbf{B}^{-1} = \mathbf{B}^{\mathbf{H}} = (\mathbf{B}^*)^{\mathbf{T}}$, where $\mathbf{B}^{\mathbf{H}}$ denotes the *Hermitian transpose* of **B**.

## 2.2.1 Discrete Fourier Transform (DFT)

In *discrete Fourier transform* (DFT) [45, 100] of an $N$-dimensional vector, basis vectors are given by

$$b_k(n) = \frac{1}{\sqrt{N}}e^{j2\pi \frac{k}{N}n}, \text{ for } 0 \le n \le N-1, \text{ and } 0 \le k \le N-1. \tag{2.47}$$

DFT of $f(n), 0 \le n \le N-1$ is defined by the following expression:

$$\mathbb{F}(f(n)) = \hat{f}(k) = \sum_{n=0}^{N-1} f(n)e^{-j2\pi \frac{k}{N}n} \text{ for } 0 \le k \le N-1. \tag{2.48}$$

Following our earlier convention (see Section 2.1.3), we denote the DFT of $f(n)$ as either $\mathbb{F}(f(n))$ or $\hat{f}(k)$. The basis vectors in Eq. (2.48) are $\sqrt{N}$ times those of orthonormal basis vectors (in Eq. (2.47)). Hence the inverse DFT (IDFT) is obtained from $\hat{f}(k)$'s by the following computation.

$$f(n) = \frac{1}{N} \sum_{k=0}^{N-1} \hat{f}(k)e^{j2\pi \frac{k}{N}n} \text{ for } 0 \le n \le N-1. \tag{2.49}$$

(a)                                        (b)

**Figure 2.1**: An example of periodic extension: (a) a finite length sequence and (b) its periodic extension.

### 2.2.1.1   The Transform Matrix

Let the sequence $x(n), 0 \leq n \leq N - 1$ be represented by the column vector $\mathbf{x}$. The DFT of $\mathbf{x}$ can be expressed also by the following matrix operation:

$$\mathbf{X} = \mathbf{Fx}, \tag{2.50}$$

where the $N \times N$ transform matrix $\mathbf{F}$ is given by

$$\mathbf{F} = \left[ e^{-j2\pi \frac{k}{N} n} \right]_{0 \leq (k,n) \leq N-1}. \tag{2.51}$$

From the expression of the inverse DFT, we may easily prove that

$$\mathbf{F}^{-1} = \frac{1}{N} \mathbf{F}^{H}, \tag{2.52}$$

where $\mathbf{F}^{H}$ is the Hermitian transpose of $\mathbf{F}$.

### 2.2.1.2   Discrete Fourier Transform as Fourier Series of a Periodic Function

We may consider the DFT of $f(n), 0 \leq n \leq N - 1$ as the Fourier series of the periodic discrete function with a period of $N X_0$ where the sampling interval is $X_0$. A typical example of a sequence of length 4 is shown in Figure 2.1(b). The definition of one period $f_p(x)$ is given as

$$f_p(x) = \begin{cases} f(x), & x = 0, X_0, 2X_0, ..., (N - 1)X_0, \\ 0, & \text{otherwise.} \end{cases} \tag{2.53}$$

With the foregoing definition, $f(n)$ becomes a periodic sequence of period $N$ such that $f(n + N) = f(n)$. In that case, the *fundamental radian frequency* becomes $\omega_0 = 2\pi \frac{1}{N X_0}$, and the *Fourier series* consists of components of its *harmonics*, say $\frac{k}{NX_0}$ for $k = 0, 1, 2, \ldots$ . They are computed as follows:

$$
\begin{aligned}
\lambda(\tfrac{k}{NX_0}) &= \frac{1}{NX_0} \sum_{n=0}^{N-1} f(nX_0) e^{-j2\pi \frac{k}{NX_0} nX_0} \Delta x, \\
&= \frac{1}{N} \sum_{n=0}^{N-1} f(nX_0) e^{-j2\pi \frac{k}{N} n} \qquad \text{(Here, } \Delta x = X_0\text{)}.
\end{aligned}
\tag{2.54}
$$

The transform coefficient $\hat{f}(k)$, DFT of $f(n)$ is $N$ times the frequency component corresponding to the frequency $\frac{k}{N}$. Hence, $\frac{k}{N}$ denotes the *normalized
frequency* (considering $X_0 = 1$), so that the *normalized radian frequency* at
$k = N$ is $2\pi$. With this observation, the following two useful properties of
Fourier series can be extended to DFT.

1. $\hat{f}(N + k) = \hat{f}(k)$. (according to Eq. (2.32)).

2. For real $f(n)$, $\hat{f}(k) = \hat{f}^*(\frac{N}{2} + k)$ ( in this case, $\hat{f}(-k) = \hat{f}^*(k)$ and
   $\hat{f}(-k) = \hat{f}(N - k)$.)

The above two properties could also be proved from the definition of DFT by
exploiting the properties of complex exponentials.

### 2.2.1.3 Circular Convolution

**Definition 2.4** *The* linear convolution *between two discrete sequences $f(n)$
and $h(n)$ is defined as the discretized version of Eq. (2.25) in the following
way:*

$$
f \bigstar h(n) = \sum_{m=-\infty}^{\infty} f(m) h(n - m).
\tag{2.55}
$$

□

**Definition 2.5** *For two periodic sequences of same period $N$, circular convolution is defined by considering $N$ functional values within a period only
and extending the periodic definition of the function to the samples outside
the base interval.*

$$
\begin{aligned}
f \circledast h(n) &= \sum_{m=0}^{N-1} f(m) h(n - m), \\
&= \sum_{m=0}^{n} f(m) h(n - m) + \sum_{m=n+1}^{N-1} f(m) h(n - m + N).
\end{aligned}
\tag{2.56}
$$

□

It can be proved that the convolution–multiplication property of the Fourier
transform holds for the circular convolution operation for DFT, as stated in
the following theorem.

Table 2.2: Properties of discrete Fourier transforms of sequences of length $N$

| Name | Given input conditions | Resulting DFT |
|---|---|---|
| Linearity | $ax(n) + by(n)$, $a$ and $b$ are arbitrary scalar constants | $a\hat{x}(k) + b\hat{y}(k)$ |
| Circular time shifting | $x(< n - n_0 >_N)$ | $e^{-j2\pi \frac{k}{N} n_0}\hat{x}(k)$ |
| Circular frequency shifting | $e^{j2\pi \frac{k_0}{N} n}x(n)$ | $\hat{x}(< k - k_0 >_N)$ |
| Duality | $\hat{x}(n)$ | $Nx(< -k >_N)$ |
| Convolution–multiplication property | $x \circledast h(n)$ | $\hat{x}(k)\hat{h}(k)$ |
| Multiplication–convolution property | $x(n)y(n)$ | $\frac{1}{N}\hat{x} \circledast \hat{h}(k)$ |

**Theorem 2.5** *Given two finite length sequences $f(n)$ and $h(n)$ and their DFTs as $\hat{f}(k)$ and $\hat{g}(k)$, respectively, the DFT of their circular convolution is the same as the product of their DFTs, that is,*

$$\mathbb{F}(f \circledast h(n)) = \hat{f}(k)\hat{h}(k). \tag{2.57}$$

□

#### 2.2.1.4   Energy Preservation

Considering the orthonormal basis vectors that are $\sqrt{N}$ times the basis vectors (see Eq. (2.47)) in the expression of DFT, we can prove the following two properties of the conservation of energy of the function due to Perseval and Plancherel [45].

$$\vec{x} \cdot \vec{y} = \frac{1}{N} \vec{\hat{x}} \cdot \vec{\hat{y}}, \tag{2.58}$$

$$||x||^2 = \vec{x} \cdot \vec{x} = \frac{1}{N} \vec{\hat{x}} \cdot \vec{\hat{x}} = \frac{1}{N}||\hat{x}||^2. \tag{2.59}$$

Please note that following similar convention of denoting the DFT of a sequence in this text, the DFT of a vector $\vec{x}$ is expressed in the above as $\vec{\hat{x}}$.

#### 2.2.1.5   Other Properties

Other properties remain the same as those of the Fourier transform by considering the translation in the functional domain or frequency domain as the *circular shift* of the sequences. A circular shift of $M$ samples in the sequence $f(n)$ makes it a sequence as $f((n - M) \bmod N)$. Let the operation $x \bmod N$ be denoted as $< x >_N$. In Table 2.2, the different properties of DFT of sequences of length $N$ [100] are listed.

Table 2.3: Different types of DFTs of a sequence $f(n)$

| $\alpha$ | $\beta$ | Transform name | Notation |
|---|---|---|---|
| 0 | 0 | Discrete Fourier Transform ($DFT$) | $\hat{f}(k)$ |
| 0 | $\frac{1}{2}$ | Odd Time Discrete Fourier Transform ($OTDFT$) | $\hat{f}_{0,\frac{1}{2}}(k)$ |
| $\frac{1}{2}$ | 0 | Odd Frequency Discrete Fourier Transform ($OFDFT$) | $\hat{f}_{\frac{1}{2},0}(k)$ |
| $\frac{1}{2}$ | $\frac{1}{2}$ | Odd Frequency Odd Time Discrete Fourier Transform ($O^2DFT$) | $\hat{f}_{\frac{1}{2},\frac{1}{2}}(k)$ |

## 2.2.2 Generalized Discrete Fourier Transform (GDFT)

The basis vectors used for discrete Fourier transform may be considered as a special case of a generalized form where the sampled points of exponential harmonics may have a constant shift of $\beta$ in its domain, as well as the harmonic frequencies also being shifted by a constant $\alpha$. Such a general form of the orthonormal set of basis vectors is given below:

$$b_k^{(\alpha,\beta)}(n) = \frac{1}{\sqrt{N}} e^{j2\pi \frac{k+\alpha}{N}(n+\beta)}, \text{ for } 0 \leq n \leq N-1, \text{ and } 0 \leq k \leq N-1. \quad (2.60)$$

As before, the GDFT [95] is defined as follows:

$$\mathbb{F}_{\alpha,\beta}(f(n)) = \hat{f}_{\alpha,\beta}(k) = \sum_{n=0}^{N-1} f(n)e^{-j2\pi \frac{k+\alpha}{N}(n+\beta)}, \text{ for } 0 \leq k \leq N-1. \quad (2.61)$$

Similar to DFT in our notation we denote the GDFT of $f(n)$ as either $\mathbb{F}_{\alpha,\beta}(f(n))$ or $\hat{f}_{\alpha,\beta}(k)$. In the above expression also, the basis vectors are $\sqrt{N}$ times those of orthonormal basis vectors (in Eq. (2.60)). Accordingly, the inverse GDFT (IGDFT) is obtained from $\hat{f}_{\alpha,\beta}(k)$'s by the following computation.

$$f(n) = \frac{1}{N} \sum_{k=0}^{N-1} \hat{f}_{\alpha,\beta}(k)e^{j2\pi \frac{k+\alpha}{N}(n+\beta)}, \text{ for } 0 \leq n \leq N-1. \quad (2.62)$$

DFT is a special case of GDFT where $\alpha = \beta = 0$. Other relevant examples (including DFT) are listed in Table 2.3 [95]. Two useful properties [95] are stated below in the form of theorems reflecting relationships among DFT, OFDFT, and OTDFT. For definitions of OFDFT and OTDFT, see Table 2.3.

**Theorem 2.6** *Let $\hat{x}_{\frac{1}{2},0}(k), 0 \leq k \leq N-1$ be the OFDFT of a sequence $x(n)$ of length $N$. If $x(n)$ is padded with $N$ zeroes to make it a sequence $x'(n)$ of length $2N$, an odd coefficient $\hat{x}_{0,0}'(2k+1)$ of its $2N$ length DFT is the same as the $k$-th coefficient $\hat{x}_{\frac{1}{2},0}(k)$ of the $N$-length ODFT of $x(n)$ , that is,*

$$\hat{x}_{0,0}'(2k+1) = \hat{x}_{\frac{1}{2},0}(k), \text{ for } 0 \leq k \leq N-1. \quad (2.63)$$

☐

**Theorem 2.7** *Let* $\hat{x}_{0,\frac{1}{2}}(k), 0 \le k \le N-1$ *be the OTDFT of a sequence* $x(n)$ *of length* $N$. *If* $\hat{x}_{0,\frac{1}{2}}(k)$ *is padded with* $N$ *zeroes to make it a sequence* $\hat{x}'_{0,\frac{1}{2}}(k), 0 \le k \le 2N-1$ *of length* $2N$ *in the transform space, every odd sample corresponding to its* $2N$ *length inverse (IOTDFT)* $x'(2n+1)$ *is half the n-th sample* $x(n)$ *of the given sequence , that is,*

$$x'(2n+1) = \frac{x(n)}{2}, \ for \ 0 \le n \le N-1. \tag{2.64}$$

☐

### 2.2.2.1   Transform Matrices

The transform matrix $\mathbf{F}_{\alpha,\beta}$ for the GDFT is given in the following form:

$$\mathbf{F}_{\alpha,\beta} = \left[ e^{-j2\pi \frac{k+\alpha}{N}(n+\beta)} \right]_{0 \le (k,n) \le N-1}. \tag{2.65}$$

The following relationships hold among their inverses:

$$\begin{aligned}
\mathbf{F}_{0,0}^{-1} &= \tfrac{1}{N}\mathbf{F}_{0,0}^{H} &= \tfrac{1}{N}\mathbf{F}_{0,0}^{*}, \\
\mathbf{F}_{\frac{1}{2},0}^{-1} &= \tfrac{1}{N}\mathbf{F}_{\frac{1}{2},0}^{H} &= \tfrac{1}{N}\mathbf{F}_{0,\frac{1}{2}}^{*}, \\
\mathbf{F}_{0,\frac{1}{2}}^{-1} &= \tfrac{1}{N}\mathbf{F}_{0,\frac{1}{2}}^{H} &= \tfrac{1}{N}\mathbf{F}_{\frac{1}{2},0}^{*}, and \\
\mathbf{F}_{\frac{1}{2},\frac{1}{2}}^{-1} &= \tfrac{1}{N}\mathbf{F}_{\frac{1}{2},\frac{1}{2}}^{H} &= \tfrac{1}{N}\mathbf{F}_{\frac{1}{2},\frac{1}{2}}^{*}.
\end{aligned} \tag{2.66}$$

### 2.2.2.2   Convolution–Multiplication Properties

Let us consider how convolution–multiplication properties [95] are further generalized in the case of GDFTs. Previously, we have discussed that multiplication in the DFT domain is equivalent to circular convolution in the original domain of the function. Though the same property remains valid for some of the GDFTs (e.g., DFT itself is a special type of GDFT), this is not true for other GDFTs. For those GDFTs, this convolution operation may need to be defined differently, and with this modified definition, the multiplication operation in the transform domain becomes equivalent to it. This type of convolution is known as *skew circular convolution*, which is defined through the antiperiodic extension of functions.

**Definition 2.6** *A function* $f(x)$ *is said to be* antiperiodic *with a period* $N$ *if* $f(x+N) = -f(x)$. ☐

We consider a *general periodic* function with a period $N$ to be either *periodic* or *antiperiodic*. Hence, the *strict period* of a *general periodic* function depends on whether it is *periodic* or *antiperiodic*. For the former type, it remains the same as its period; however, for the latter, it is *twice* of its *period*.

**Definition 2.7** The skew circular convolution *for two sequences of the same length N is defined as linear convolution of their antiperiodic extensions of a period N, and the resulting sequence is formed by N functional values within its period only.*

$$
\begin{aligned}
f \textcircled{S} h(n) &= \sum_{m=0}^{N-1} f(m)h(n-m), \\
&= \sum_{m=0}^{n} f(m)h(n-m) - \sum_{m=n+1}^{N-1} f(m)h(n-m+N).
\end{aligned}
$$

(2.67)

$\square$

With the definitions of circular and skewcircular convolutions, the convolution multiplication properties of different GDFTs are stated by the following theorem [95].

**Theorem 2.8** *Given two finite length sequences $f(n)$ and $h(n)$ and their GDFTs as $\hat{f}_{\alpha,\beta}(k)$ and $\hat{g}_{\alpha,\beta}(k)$, respectively, the following convolution multiplication properties are observed,*

$$
\begin{aligned}
\mathbb{F}_{0,0}(f \circledast h(n)) &= \hat{f}_{0,0}(k)\hat{h}_{0,0}(k), \\
\mathbb{F}_{0,\frac{1}{2}}(f \circledast h(n)) &= \hat{f}_{0,\frac{1}{2}}(k)\hat{h}_{0,0}(k), \\
\mathbb{F}_{0,0}(f \circledast h(n-1)) &= \hat{f}_{0,\frac{1}{2}}(k)\hat{h}_{0,\frac{1}{2}}(k), \\
\mathbb{F}_{\frac{1}{2},0}(f \textcircled{S} h(n)) &= \hat{f}_{\frac{1}{2},0}(k)\hat{h}_{\frac{1}{2},0}(k), \\
\mathbb{F}_{\frac{1}{2},\frac{1}{2}}(f \textcircled{S} h(n)) &= \hat{f}_{\frac{1}{2},\frac{1}{2}}(k)\hat{h}_{\frac{1}{2},0}(k), \; and, \\
\mathbb{F}_{\frac{1}{2},0}(f \textcircled{S} h(n-1)) &= \hat{f}_{\frac{1}{2},\frac{1}{2}}(k)\hat{h}_{\frac{1}{2},\frac{1}{2}}(k).
\end{aligned}
$$

(2.68)

$\square$

### 2.2.3 Discrete Trigonometric Transforms

In Section 2.1.2, it was pointed out that for an *even periodic* function, orthogonal expansion is possible with only *cosine harmonics*, and for an *odd periodic* function, its *Fourier series expansion* requires only *sine harmonics* as basis functions. These properties are suitably exploited for a finite discrete sequence to have orthogonal transforms with only *discrete cosine* or *discrete sine* basis vectors. These transforms are defined under the same framework of GDFT, by symmetrically (or antisymmetrically) extending the sequence at both its ends so that the extended sequence becomes either *even* or *odd* sequence about a point of symmetry. Hence, the GDFT over the extended sequence becomes equivalent to Fourier series of an *even* or *odd* periodic function. Thus, all these expansions can be expressed by a linear combination of either discrete cosine [3] or discrete sine basis vectors. These transforms in general are referred to as *discrete trigonometric transforms*. To understand the different types of trigonometric transforms, let us first look at the various types of symmetric extensions of a finite length sequence.

### 2.2.3.1   Symmetric Extensions of Finite Sequences

Consider a finite length sequence $x(n), 0 \leq n \leq N-1$ of length $N$. The sequence may be symmetrically extended at its end point (say around the last sample point $x(N-1)$) in *two* different ways. The extended sample is *symmetric* about the sample point itself. We refer to this type of *symmetric extension* as the *whole-sample symmetric (WS) extension*. Alternatively, the symmetric extension may be made centering about the midpoint between the $(N-1)$th and $N$th sample. This type of extension is referred to as *half-sample symmetric (HS) extension*. Similarly, there are *two* types of *antisymmetric extensions* about a sample point of a sequence. They are referred to as *whole-sample antisymmetric (WA) extension* and *half-sample antisymmetric extension*, respectively. In Figure 2.2, all these extensions about the right end sample point of a finite length sequence of length 4 are illustrated. Note that for a *WA* extension, a zero has to be introduced at the $N$th sample point. These extensions are mathematically defined below:

1. For the WS extension at the end sample point,

$$\tilde{x}(n) = \begin{cases} x(n) & 0 \leq n \leq N-1 \\ x(2N-n-2) & N \leq n \leq 2N-2 \end{cases} \qquad (2.69)$$

2. For the HS extension at the end sample point,

$$\tilde{x}(n) = \begin{cases} x(n) & 0 \leq n \leq N-1 \\ x(2N-n-1) & N \leq n \leq 2N-1 \end{cases} \qquad (2.70)$$

3. For the WA extension at the end sample point,

$$\tilde{x}(n) = \begin{cases} x(n) & 0 \leq n \leq N-1 \\ 0 & n = N \\ -x(2N-n) & N+1 \leq n \leq 2N \end{cases} \qquad (2.71)$$

4. For the HA extension at the end sample point,

$$\tilde{x}(n) = \begin{cases} x(n) & 0 \leq n \leq N-1 \\ -x(2N-n-1) & N+1 \leq n \leq 2N-1 \end{cases} \qquad (2.72)$$

### 2.2.3.2   Symmetric Periodic Extension

Depending upon the nature of periodic symmetric extension at the two ends (at sample positions 0 and $N-1$) of a finite length sequence of length $N$, it becomes either a periodic or an antiperiodic sequence. The resulting period of the extended sequence depends upon types of extensions made at its two ends. Moreover, a strict period of a sequence becomes either *even* or *odd*. If the symmetric extension at the left end (the 0th sample position) is either

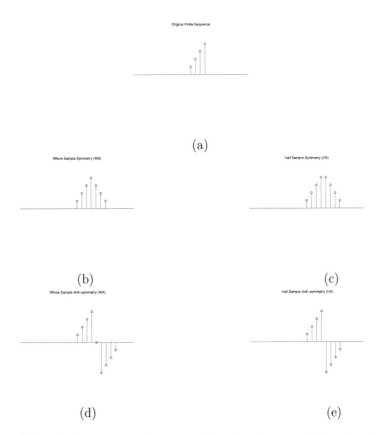

**Figure 2.2**: Different types of symmetric and antisymmetric extensions at the end of a sequence: (a) Original, (b) WS, (c) HS, (d) WA, and (e) HA.

**Table 2.4**: Summary of observations of symmetric periodic extension of a sequence of length $N$

| Symmetric extensions | Periodic (P) / antiperiodic (A) | Length of period | Even (E) or Odd (O) |
|---|---|---|---|
| WSWS | P | $2N - 2$ | E |
| WSHS | P | $2N - 1$ | E |
| HSHS | P | $2N$ | E |
| HSWS | P | $2N - 1$ | E |
| WSWA | A | $2N$ | E |
| WSHA | A | $2N - 1$ | E |
| HSHA | A | $2N$ | E |
| HSWA | A | $2N - 1$ | E |
| WAWS | A | $2N$ | O |
| WAHS | A | $2N + 1$ | O |
| HAHS | A | $2N$ | O |
| HAWS | A | $2N - 1$ | O |
| WAWA | P | $2N + 2$ | O |
| WAHA | P | $2N + 1$ | O |
| HAHA | P | $2N$ | O |
| HAWA | P | $2N + 1$ | O |

WS or HS, the strict period becomes even; otherwise it is odd. Therefore their expansions with either a set of cosine or sine basis vectors are possible. In Figures 2.3 and 2.4, examples of such strict periods of symmetric periodic extension of a finite length sequence of length 4 are shown. The corresponding extensions are encoded by the successive short form of the extensions (WS, HS, WA, or HA) with the first two letters for the left end and the last two letters for the right end of the sequence, respectively. Samples of the original sequence are marked by a double arrow under the horizontal axis. We may observe that the resulting sequences of length $N$ in *general periodic* sense become either periodic or antiperiodic. A summary of observations is provided in Table 2.4.

Each type of symmetric extension in the above table is associated with an appropriate GDFT, which eventually provides us a type of discrete cosine transform or discrete sine transform. The length of the original sequence is chosen in such a manner ($N - 1$, $N$ or $N + 1$) that the length of the *general period* becomes either $2N$ or $2N-1$. Considering the parity of the length of this period, they are termed as *even* or *odd trigonometric transform*. In Tables 2.5 and 2.6 [95] different types of DCTs and DSTs and the corresponding GDFTs deriving them are provided. Corresponding functional forms of symmetric extensions are also shown in those tables. Expressions of these transforms are given in the following subsection.

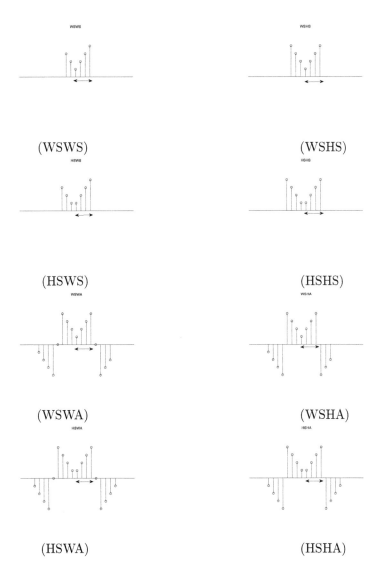

**Figure 2.3**: Symmetrically extended even periodic sequences from a finite length sequence of length 4.

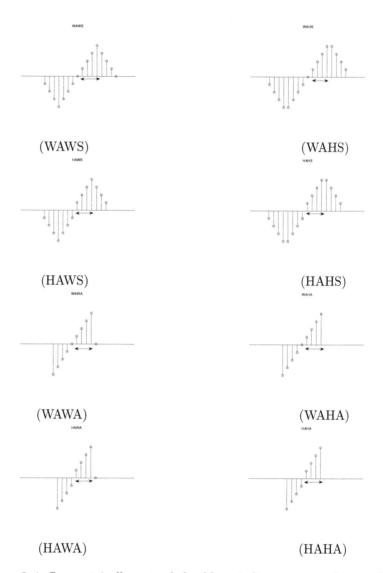

(WAWS)  (WAHS)

(HAWS)  (HAHS)

(WAWA)  (WAHA)

(HAWA)  (HAHA)

**Figure 2.4**: Symmetrically extended odd periodic sequences from a finite length sequence of length 4.

Table 2.5: Different types of discrete cosine transforms (DCT)

| Length of input | Symmetric extensions: $\tilde{x}(n) = \mathcal{E}(x(n))$ | Length of period | GDFT | DCT type |
|---|---|---|---|---|
| $N+1$ | WSWS: $\tilde{x}(n) = \begin{cases} x(n), & n = 0,1,2,\ldots,N, \\ x(2N-n), & n = N+1,\ldots,2N-1. \end{cases}$ | $2N$ | $\mathbf{F}_{0,0}$ | $C_{1e}$ |
| $N$ | WSHS: $\tilde{x}(n) = \begin{cases} x(n), & n = 0,1,2,\ldots,N-1, \\ x(2N-1-n), & n = N,\ldots,2N-2. \end{cases}$ | $2N-1$ | $\mathbf{F}_{0,0}$ | $C_{1o}$ |
| $N$ | HSHS: $\tilde{x}(n) = \begin{cases} x(n), & n = 0,1,2,\ldots,N-1, \\ x(2N-1-n), & n = N,\ldots,2N-1. \end{cases}$ | $2N$ | $\mathbf{F}_{0,\frac{1}{2}}$ | $C_{2e}$ |
| $N$ | HSWS: $\tilde{x}(n) = \begin{cases} x(n), & n = 0,1,2,\ldots,N-1, \\ x(2N-2-n), & n = N,\ldots,2N-2. \end{cases}$ | $2N-1$ | $\mathbf{F}_{0,\frac{1}{2}}$ | $C_{2o}$ |
| $N$ | WSWA: $\tilde{x}(n) = \begin{cases} x(n), & n = 0,1,2,\ldots,N-1, \\ 0, & n = N, \\ -x(2N-n), & n = N+1,\ldots,2N-1. \end{cases}$ | $2N$ | $\mathbf{F}_{\frac{1}{2},0}$ | $C_{3e}$ |
| $N$ | WSHA: $\tilde{x}(n) = \begin{cases} x(n), & n = 0,1,2,\ldots,N-1, \\ -x(2N-1-n), & n = N,\ldots,2N-2. \end{cases}$ | $2N-1$ | $\mathbf{F}_{\frac{1}{2},0}$ | $C_{3o}$ |
| $N$ | HSHA: $\tilde{x}(n) = \begin{cases} x(n), & n = 0,1,2,\ldots,N-1, \\ -x(2N-1-n), & n = N,\ldots,2N-1. \end{cases}$ | $2N$ | $\mathbf{F}_{\frac{1}{2},\frac{1}{2}}$ | $C_{4e}$ |
| $N$ | HSWA: $\tilde{x}(n) = \begin{cases} x(n), & n = 0,1,2,\ldots,N-2, \\ 0, & n = N-1, \\ -x(2N-2-n), & n = N,\ldots,2N-2. \end{cases}$ | $2N-1$ | $\mathbf{F}_{\frac{1}{2},\frac{1}{2}}$ | $C_{4o}$ |

Table 2.6: Different types of discrete sine transforms (DST)

| Length of input | Symmetric extensions: $\tilde{x}(n) = \mathcal{E}(x(n))$ | Length of period | GDFT | DST type |
|---|---|---|---|---|
| $N-1$ | WAWA: $\tilde{x}(n) = \begin{cases} 0, & n = 0, \\ x(n), & n = 1,2,\ldots,N-1, \\ 0, & n = N, \\ -x(2N-n), & n = N+1,\ldots,2N-1. \end{cases}$ | $2N$ | $j\mathbf{F}_{0,0}$ | $S_{1e}$ |
| $N-1$ | WAHA: $\tilde{x}(n) = \begin{cases} 0, & n = 0, \\ x(n), & n = 0,1,2,\ldots,N-1, \\ -x(2N-1-n), & n = N,\ldots,2N-2. \end{cases}$ | $2N-1$ | $j\mathbf{F}_{0,0}$ | $S_{1o}$ |
| $N$ | HAHA: $\tilde{x}(n) = \begin{cases} x(n), & n = 0,1,2,\ldots,N-1, \\ -x(2N-1-n), & n = N,\ldots,2N-1. \end{cases}$ | $2N$ | $j\mathbf{F}_{0,\frac{1}{2}}$ | $S_{2e}$ |
| $N-1$ | HAWA: $\tilde{x}(n) = \begin{cases} x(n), & n = 0,1,2,\ldots,N-2, \\ 0, & n = N-1, \\ -x(2N-2-n), & n = N,\ldots,2N-2. \end{cases}$ | $2N-1$ | $j\mathbf{F}_{0,\frac{1}{2}}$ | $S_{2o}$ |
| $N$ | WAWS: $\tilde{x}(n) = \begin{cases} 0, & n = 0, \\ x(n), & n = 1,2,\ldots,N, \\ -x(2N-n), & n = N,\ldots,2N-1. \end{cases}$ | $2N$ | $j\mathbf{F}_{\frac{1}{2},0}$ | $S_{3e}$ |
| $N-1$ | WAHS: $\tilde{x}(n) = \begin{cases} 0, & n = 0, \\ x(n), & n = 1,2,\ldots,N-1, \\ x(2N-1-n), & n = N,\ldots,2N-2. \end{cases}$ | $2N-1$ | $j\mathbf{F}_{\frac{1}{2},0}$ | $S_{3o}$ |
| $N$ | HAHS: $\tilde{x}(n) = \begin{cases} x(n), & n = 0,1,2,\ldots,N-1, \\ x(2N-1-n), & n = N,\ldots,2N-1. \end{cases}$ | $2N$ | $j\mathbf{F}_{\frac{1}{2},\frac{1}{2}}$ | $S_{4e}$ |
| $N$ | HAWS: $\tilde{x}(n) = \begin{cases} x(n), & n = 0,1,2,\ldots,N-1, \\ x(2N-1-n), & n = N,\ldots,2N-2. \end{cases}$ | $2N-1$ | $j\mathbf{F}_{\frac{1}{2},\frac{1}{2}}$ | $S_{4o}$ |

### 2.2.3.3 Different Types of Discrete Trigonometric Transforms

Let $x(n), n = 0, 1, 2, \ldots$ , be a sequence of input data. Then different types of $N$−point DCTs and DSTs are defined as follows:

Type-I Even DCT:

$$C_{1e}(x(n)) = X_{Ie}(k) = \sqrt{\frac{2}{N}} \alpha^2(k) \sum_{n=0}^{N} x(n) \cos\left(\frac{2\pi nk}{2N}\right), \ 0 \leq k \leq N, \quad (2.73)$$

Type-I Odd DCT:

$$C_{1o}(x(n)) = X_{Io}(k) = \sqrt{\frac{2}{N}} \alpha^2(k) \sum_{n=0}^{N-1} x(n) \cos\left(\frac{2\pi nk}{2N-1}\right), \ 0 \leq k \leq N-1, \quad (2.74)$$

Type-II Even DCT:

$$C_{2e}(x(n)) = X_{IIe}(k) = \sqrt{\frac{2}{N}} \alpha(k) \sum_{n=0}^{N-1} x(n) \cos\left(\frac{2\pi k(n+\frac{1}{2})}{2N}\right), \\ 0 \leq k \leq N-1, \quad (2.75)$$

Type-II Odd DCT:

$$C_{2o}(x(n)) = X_{IIo}(k) = \sqrt{\frac{2}{N}} \alpha(k)\beta(k) \sum_{n=0}^{N-1} x(n) \cos\left(\frac{2\pi k(n+\frac{1}{2})}{2N-1}\right), \\ 0 \leq k \leq N-1, \quad (2.76)$$

Type-III Even DCT:

$$C_{3e}(x(n)) = X_{IIIe}(k) = \sqrt{\frac{2}{N}} \alpha(k) \sum_{n=0}^{N-1} x(n) \cos\left(\frac{2\pi(k+\frac{1}{2})n}{2N}\right), \\ 0 \leq k \leq N-1, \quad (2.77)$$

Type-III Odd DCT:

$$C_{3o}(x(n)) = X_{IIIo}(k) = \sqrt{\frac{2}{N}} \alpha(k)\beta(k) \sum_{n=0}^{N-1} x(n) \cos\left(\frac{2\pi(k+\frac{1}{2})n}{2N-1}\right), \\ 0 \leq k \leq N-1, \quad (2.78)$$

Type-IV Even DCT:

$$C_{4e}(x(n)) = X_{IVe}(k) = \sqrt{\frac{2}{N}} \sum_{n=0}^{N-1} x(n) \cos\left(\frac{2\pi(k+\frac{1}{2})(n+\frac{1}{2})}{2N}\right),$$
$$0 \le k \le N-1, \quad (2.79)$$

Type-IV Odd DCT:

$$C_{4o}(x(n)) = X_{IVo}(k) = \sqrt{\frac{2}{N}} \sum_{n=0}^{N-2} x(n) \cos\left(\frac{2\pi(k+\frac{1}{2})(n+\frac{1}{2})}{2N-1}\right),$$
$$0 \le k \le N-2, \quad (2.80)$$

Type-I Even DST:

$$S_{1e}(x(n)) = X_{sIe}(k) = \sqrt{\frac{2}{N}} \sum_{n=1}^{N-1} x(n) \sin\left(\frac{2\pi kn}{2N}\right), \ 1 \le k \le N-1, \quad (2.81)$$

Type-I Odd DST:

$$S_{1o}(x(n)) = X_{sIo}(k) = \sqrt{\frac{2}{N}} \sum_{n=1}^{N-1} x(n) \sin\left(\frac{2\pi kn}{2N-1}\right), \ 1 \le k \le N-1, \quad (2.82)$$

Type-II Even DST:

$$S_{2e}(x(n)) = X_{sIIe}(k) = \sqrt{\frac{2}{N}}\alpha(k) \sum_{n=0}^{N-1} x(n) \sin\left(\frac{2\pi k(n+\frac{1}{2})}{2N}\right),$$
$$1 \le k \le N-1, \quad (2.83)$$

Type-II Odd DST:

$$S_{2o}(x(n)) = X_{sIIo}(k) = \sqrt{\frac{2}{N}} \sum_{n=0}^{N-1} x(n) \sin\left(\frac{2\pi k(n+\frac{1}{2})}{2N-1}\right), \ 1 \le k \le N-1, \quad (2.84)$$

Type-III Even DST:

$$S_{3e}(x(n)) = X_{sIIIe}(k) = \sqrt{\frac{2}{N}}\alpha(k) \sum_{n=1}^{N} x(n) \sin\left(\frac{2\pi(k+\frac{1}{2})(n)}{2N}\right),$$
$$0 \le k \le N-1, \quad (2.85)$$

Type-III Odd DST:

$$S_{3o}(x(n)) = X_{sIIIo}(k) = \sqrt{\frac{2}{N}} \sum_{n=1}^{N} x(n) \sin\left(\frac{2\pi(k+\frac{1}{2})n}{2N-1}\right), \quad 0 \le k \le N-2,$$

(2.86)

Type-IV Even DST:

$$S_{4e}(x(n)) = X_{sIVe}(k) = \sqrt{\frac{2}{N}} \sum_{n=0}^{N-1} x(n) \sin\left(\frac{2\pi(k+\frac{1}{2})(n+\frac{1}{2})}{2N}\right),$$
$$0 \le k \le N-1,$$

(2.87)

Type-IV Odd DST:

$$S_{4o}(x(n)) = X_{sIVo}(k) = \sqrt{\frac{2}{N}} \beta^2(k) \sum_{n=0}^{N-1} x(n) \sin\left(\frac{2\pi(k+\frac{1}{2})(n+\frac{1}{2})}{2N-1}\right),$$
$$0 \le k \le N-1,$$

(2.88)

where $\alpha(p)$ and $\beta(p)$ in the foregoing equations are given by

$$\alpha(p) = \begin{cases} \sqrt{\frac{1}{2}}, & \text{for } p = 0 \text{ or } N, \\ 1, & \text{otherwise.} \end{cases}$$

(2.89)

$$\beta(p) = \begin{cases} \sqrt{\frac{1}{2}}, & \text{for } p = N-1, \\ 1, & \text{otherwise.} \end{cases}$$

(2.90)

All these transforms are linear, distributive, and invertible. The inverse relationships between the transform domain and the original signal domain are stated below [95]:

$$\begin{aligned}
x(n) &= C_{1e}\{C_{1e}\{x(n)\}\}, \\
x(n) &= C_{3e}\{C_{2e}\{x(n)\}\}, \\
x(n) &= C_{2e}\{C_{3e}\{x(n)\}\}, \\
x(n) &= C_{4e}\{C_{4e}\{x(n)\}\}. \\
x(n) &= C_{1o}\{C_{1o}\{x(n)\}\}, \\
x(n) &= C_{3o}\{C_{2o}\{x(n)\}\}, \\
x(n) &= C_{2o}\{C_{3o}\{x(n)\}\}, \\
x(n) &= C_{4o}\{C_{4o}\{x(n)\}\}. \\
x(n) &= S_{1e}\{S_{1e}\{x(n)\}\}, \\
x(n) &= S_{3e}\{S_{2e}\{x(n)\}\}, \\
x(n) &= S_{2e}\{S_{3e}\{x(n)\}\}, \\
x(n) &= S_{4e}\{S_{4e}\{x(n)\}\}. \\
x(n) &= S_{1o}\{S_{1o}\{x(n)\}\}, \\
x(n) &= S_{3o}\{S_{2o}\{x(n)\}\}, \\
x(n) &= S_{2o}\{S_{3o}\{x(n)\}\}, \\
x(n) &= S_{4o}\{S_{4o}\{x(n)\}\}.
\end{aligned}$$

(2.91)

In a DCT-based compression algorithm, *type-II even DCT* is used for transforming discrete functions. In our subsequent discussion, we concentrate on *even* DCTs and DSTs only, and all four types of even DCTs and DSTs are simply referred to as *type-I, type-II, type-III*, and *type-IV* DCTs and DSTs, respectively. Moreover, the unqualified term DCT is to be implied as the *type-II even DCT*.

### 2.2.3.4 Convolution Multiplication Properties

The operation for which the convolution multiplication properties hold for trigonometric transforms is discrete convolution with *symmetrically extended sequences*. Depending upon the nature of the periodicity of these general periodic sequences, this operation becomes either a *circular convolution* or a *skew-circular convolution*. When both the extended sequences become periodic, *symmetric convolution* turns out to be the *circular* one. On the other hand, both being *antiperiodic* it becomes the same as the skew-circular convolution. Convolution between antiperiodic and periodic functions is not defined within its general period. This type of convolution is known as symmetric convolution, which is formally defined below [95].

**Definition 2.8** *Symmetric Convolution between two sequences $x(n)$ and $y(n)$ is defined as the convolution between their symmetrically extended sequences of* the same general period. *When both the extended sequences are* periodic, *the convolution operation is defined as the* circular convolution. *If both are antiperiodic, the operation is defined as the* skew-circular *convolution. Symmetric convolution for a periodic (in its strict sense) and antiperiodic sequence of the same general period is not defined.*

*Let the symmetric extensions of $x(n)$ and $y(n)$ of the* same general period be denoted as $\mathcal{E}_a\left(x(n)\right)$ and $\mathcal{E}_b\left(y(n)\right)$, respectively. Let us also denote the symmetric convolution *operator by '✖'. So the symmetric convolution between $x(n)$ and $y(n)$ is expressed as*

$$x ✖ y(n) = \begin{cases} \mathcal{E}_a\left(x(n)\right) \circledast \mathcal{E}_b\left(y(n)\right) & \text{if } \mathcal{E}_a(.) \text{ and } \mathcal{E}_b(.) \text{ are periodic.} \\ \mathcal{E}_a\left(x(n)\right) \circledS \mathcal{E}_b\left(y(n)\right) & \text{if } \mathcal{E}_a(.) \text{ and } \mathcal{E}_b(.) \text{ are antiperiodic.} \end{cases}$$
(2.92)

$\square$

With this defintion, the convolution multiplication properties of these transforms can be expressed in a general form as given below [95].

$$T_c\left(x ✖ y(n - n_0)\right) = \sqrt{2N} T_a\left(x(n)\right) T_b\left(y(n)\right)$$
(2.93)

where $T_a(.)$ and $T_b(.)$ are *discrete trigonometric transforms* (DTT) associated with corresponding extensions $\mathcal{E}_a(.)$ and $\mathcal{E}_b(.)$ for sequences $x(n)$ and $y(n)$, respectively, and accordingly the *symmetric convolution* operation is performed on them. $T_c(.)$ is another DTT, and $n_0$ is a constant integer whose value is either 0 or 1. Following the convolution multiplication properties of GDFT (see Eq. (2.68)), a few specific cases involving even DCTs are illustrated in the following theorem.

**Theorem 2.9** *Let $u(n) = x(n) \circledast y(n)$, and $w(n) = x(n) \circledS y(n)$. The numbers of samples in $x(n)$ and $y(n)$ are such that their symmetric extensions produce a general period of $2N$. The convolution multiplication properties involving the even DCTs are listed below [95].*

$$
\begin{aligned}
C_{1e}\left(u(n)\right) &= \sqrt{2N}C_{1e}\left(x(l)\right)C_{1e}\left(y(m)\right), && 0 \le n,l,m \le N, \\
C_{2e}\left(u(n)\right) &= \sqrt{2N}C_{2e}\left(x(l)\right)C_{1e}\left(y(m)\right), && 0 \le n,l \le N-1, \\
&&& 0 \le m \le N, \\
C_{1e}\left(u(n-1)\right) &= \sqrt{2N}C_{2e}\left(x(l)\right)C_{2e}\left(y(m)\right), && 0 \le l,m \le N-1, \\
&&& -1 \le n \le N-1, \\
C_{3e}\left(w(n)\right) &= \sqrt{2N}C_{3e}\left(x(l)\right)C_{3e}\left(y(m)\right), && 0 \le n,l,m \le N-1, \\
C_{4e}\left(w(n)\right) &= \sqrt{2N}C_{4e}\left(x(l)\right)C_{3e}\left(y(m)\right), && 0 \le n,m,l \le N-1, \text{ and,} \\
C_{3e}\left(w(n-1)\right) &= \sqrt{2N}C_{4e}\left(x(l)\right)C_{4e}\left(y(m)\right), && 0 \le l,m \le N-1, \\
&&& -1 \le n \le N-2.
\end{aligned}
\tag{2.94}
$$

□

The convolution-multiplication properties are also applicable for other different combination of trigonometric transforms. For a comprehensive list, we may refer to the paper by Martucci [95]. However, another set of relationships is provided below involving both even DCTs and DSTs at the same time.

$$
\begin{aligned}
-C_{2e}\left(u(n)\right) &= \sqrt{2N}S_{1e}\left(x(l)\right)S_{2e}\left(y(m)\right), && 1 \le l \le N-1, \\
&&& 0 \le n,m \le N-1, \\
S_{1e}\left(u(n)\right) &= \sqrt{2N}C_{1e}\left(x(l)\right)S_{1e}\left(y(m)\right), && 1 \le n,m \le N-1, \\
&&& 0 \le l \le N, \\
S_{2e}\left(u(n)\right) &= \sqrt{2N}C_{2e}\left(x(l)\right)S_{1e}\left(y(m)\right), && 1 \le l,n \le N-1, \\
&&& 1 \le m \le N-1, \\
S_{2e}\left(u(n)\right) &= \sqrt{2N}S_{2e}\left(x(l)\right)C_{1e}\left(y(m)\right), && 1 \le l,n \le N-1, \\
&&& 0 \le m \le N, \\
S_{1e}\left(u(n-1)\right) &= \sqrt{2N}C_{2e}\left(x(l)\right)S_{2e}\left(y(m)\right), && 0 \le l,m \le N-1, \\
&&& 0 \le n \le N-2, \\
S_{3e}\left(w(n)\right) &= \sqrt{2N}C_{3e}\left(x(l)\right)S_{3e}\left(y(m)\right), && 0 \le l \le N-1, \\
&&& 1 \le m,n \le N, \\
S_{4e}\left(w(n)\right) &= \sqrt{2N}C_{4e}\left(x(l)\right)S_{3e}\left(y(m)\right), && 0 \le l,n \le N-1, \\
&&& 1 \le m \le N, \\
S_{4e}\left(w(n)\right) &= \sqrt{2N}S_{4e}\left(x(l)\right)C_{3e}\left(y(m)\right), && 0 \le l,m,n \le N-1, \text{ and,} \\
S_{3e}\left(w(n-1)\right) &= \sqrt{2N}C_{4e}\left(x(l)\right)S_{4e}\left(y(m)\right), && 0 \le l,m,n \le N-1.
\end{aligned}
\tag{2.95}
$$

## 2.2.4 Type-II Even DCT

As the *Type-II even DCT* (or *Type-II DCT* as referred to subsequently) is used in DCT-based compression schemes such as JPEG, MPEG, etc., we review its properties in more detail in this section.

### 2.2.4.1 Matrix Representation

As the DCTs are linear operations, they can be conveniently represented by matrix operations. Let $C_N$ denote the $N \times N$ type-II DCT matrix whose $(k,n)$th element (for $0 \leq k \leq N-1$ and $0 \leq n \leq N-1$) is given by $\sqrt{\frac{2}{N}}.\alpha(k)cos(\frac{(2n+1)\pi n}{2N})$ (see Eq. (2.75)) as represented by the following notation:

$$C_N = \left[ \sqrt{\frac{2}{N}}.\alpha(k)cos(\frac{\pi k(2n+1)}{2N}) \right]_{0 \leq (k,n) \leq N-1}. \qquad (2.96)$$

Following the above notation, general even $N$-point DCT matrices of type-I, II, III, and IV are denoted by $C_N^I, C_N^{II}, C_N^{III}$, and $C_N^{IV}$, respectively. If the type is not mentioned in the superscript, the type of the matrix should be considered as type-II. Similarly, for *even DST*s, the transform matrices are denoted by $S_N^I, S_N^{II}, S_N^{III}$, and $S_N^{IV}$ for type-I, II, III, and IV $N$-point DST, respectively. In this case also, if the type is not mentioned in the superscript such as $S_N$, it is to be considered as a type-II DST matrix.

Let the sequence $\{x(n), n = 0, 1, ... N - 1\}$ be represented by the $N$-dimensional column-vector $\mathbf{x}$. Then the type-II DCT of $\mathbf{x}$ is given as follows:

$$X = C_N.\mathbf{x}. \qquad (2.97)$$

As the type-II DCT is an orthonormal transform, the following relationship holds with its inverse:

$$C_N^{-1} = C_N^T. \qquad (2.98)$$

Rows of $C_N$ satisfy the following: properties.

**Theorem 2.10** *For even $N$, each row of the DCT matrix $C_N$ is either symmetric (for even rows) or antisymmetric (for odd rows) about its center, that is,*

$$C_N(k, N-1-n) = \begin{cases} C_N(k,n) & for \ k \ even \\ -C_N(k,n) & for \ k \ odd \end{cases} \qquad (2.99)$$

$\square$

### 2.2.4.2 Downsampling and Upsampling Properties of the DCTs

Both type-I and type-II DCTs have interesting relationship with downsampled and upsampled sequences. They are stated below [94]:

**Theorem 2.11** *If* $X_I(k) = C_{1e}(x(n)), k, n = 0, 1, 2, .. , N,$ *then* $x_d(n) = x(2n + 1) = C_{2e}^{-1} \{ \frac{X_I(k) - X_I(N-k)}{\sqrt{2}} \}$ *for* $k, n = 0, 1, 2, .... , \frac{N}{2} - 1.$

□

**Theorem 2.12** *If* $X_{II}(k) = C_{2e}(x(n)), k, n = 0, 1, 2, ... , \frac{N}{2},$ *then*
$$x_u(n) = C_{1e}^{-1} \{ \frac{X_{II}(k) - X_{II}(N-k)}{\sqrt{2}} \} \text{ for } k, n = 0, 1, 2, ... , N, \text{ where}$$

$$x_u(n) = \begin{cases} 0, & n \quad even, \\ x(\frac{n-1}{2}), & n \quad odd. \end{cases} \tag{2.100}$$

□

Theorem 2.11 is useful for decimating signals directly in the compressed domain, while Theorem 2.12 is used in interpolation.

### 2.2.4.3    Subband Relationship of the type-II DCT

Subbands are components of a sequence in its multiresolution representation. Suppose $x(n)$, $n = 0, 1, ... , N - 1$ be an $N$-point data sequence with even $N$. Let the sequence $x(n)$ be decomposed into two subbands $x_L(n)$ and $x_H(n)$ of length $\frac{N}{2}$ each as follows:

$$\begin{aligned} x_L(n) &= \tfrac{1}{2}\{x(2n) + x(2n + 1)\}, \\ x_H(n) &= \tfrac{1}{2}\{x(2n) - x(2n + 1)\}, \quad n = 0, 1, ... , \tfrac{N}{2} - 1. \end{aligned} \tag{2.101}$$

The relationship between the DCT of subbands (of $\frac{N}{2}$ point) with the original DCT (of $N$-point) of the sequence are stated in the following theorem.

**Theorem 2.13** *Let* $x_L(n)$ *and* $x_H(n)$ *be subbands of a sequence* $x(n)$ *as computed from Eq. (2.101), and let* $X_L(k) = C_{2e}(x_L(n))$ *and* $S_H(k) = S_{2e}(x_H(n))$. *Then* $X(k)$, *DCT of* $x(n)$, *from* $X_L(k)$*'s and* $S_H(k)$*'s is obtained as follows [72]:*

$$X(k) = \sqrt{2}\cos(\frac{\pi k}{2N})\overline{X_L}(k) + \sqrt{2}\sin(\frac{\pi k}{2N})\overline{S_H}(k), \ 0 \le k \le N - 1, \tag{2.102}$$

*where*

$$\overline{X_L}(k) = \begin{cases} X_L(k), & 0 \le k \le \frac{N}{2} - 1, \\ 0, & k = \frac{N}{2}, \\ -X_L(N - k), & \frac{N}{2} + 1 \le k \le N - 1, \end{cases} \tag{2.103}$$

*and*

$$\overline{S_H}(k) = \begin{cases} S_H(k), & 0 \le k \le \frac{N}{2} - 1, \\ \sqrt{2}\sum\limits_{n=0}^{\frac{N}{2}-1}(-1)^n x_H(n), & k = \frac{N}{2}, \\ S_H(N - k), & \frac{N}{2} + 1 \le k \le N - 1. \end{cases} \tag{2.104}$$

□

### 2.2.4.4 Approximate DCT Computation

If most of the energy of a signal is concentrated in the lower-frequency zone, Eq. (2.102) can be approximated as [72]

$$X(k) = \begin{cases} \sqrt{2}\cos(\frac{\pi k}{2N})\overline{X_L}(k), & k \in \{0, 1, \dots, \frac{N}{2} - 1\}, \\ 0, & \text{otherwise.} \end{cases} \tag{2.105}$$

This approximation is referred to as the *subband approximation* of DCT. It is further simplified by removing the cosine factor as shown in the following expression:

$$X(k) = \begin{cases} \sqrt{2} \cdot \overline{X_L}(k), & k \in \{0, 1, \dots, \frac{N}{2} - 1\}. \\ 0, & \text{otherwise.} \end{cases} \tag{2.106}$$

This simplified approximation is referred to as the *low-pass truncated approximation* of DCT.

### 2.2.4.5 Composition and Decomposition of the DCT Blocks

The composition and decomposition of the DCT blocks [67] are carried out by exploiting the spatial relationship of block DCTs. Let $\{x(n), n = 0, 1, \dots, MN - 1\}$ be a sequence of length $MN$. This sequence is partitioned into $M$ blocks (or subsequences) each containing $N$ data points. In the block DCT space, an $N$-point DCT is applied to each block of $N$ data points. Hence the $N$-point DCT of the $p$-th block is expressed as follows:

$$X_p^{(N)}(k) = \sqrt{\frac{2}{N}}\alpha(k)\sum_{n=0}^{N-1} x(pN + n)\cos(\frac{(2n+1)\pi k}{2N}), \tag{2.107}$$
$$0 \leq p \leq M - 1, 0 \leq k \leq N - 1.$$

On the other hand, the $MN$-point DCT of $x(n)$ is given by

$$X^{(MN)}(k) = \sqrt{\frac{2}{MN}}\alpha(k)\sum_{n=0}^{MN-1} x(n)\cos(\frac{(2n+1)\pi k}{2 \times M \times N}), \tag{2.108}$$
$$0 \leq k \leq M \times N - 1.$$

In both of the above two equations, $\alpha(k)$ is given by Eq. (2.89).

We may observe that the block DCT transformation as expressed in Eq. (2.107) is an orthonormal expansion of the sequence $\{x(n)\}$ with a set of $MN$ basis vectors [67], each of which could be derived from the basis vectors of the $N$-point DCT by translating the basis vector with the multiple of block sizes and padding with zeroes. As this derived set is also orthonormal, there exists an invertible linear transformation from one transform to the other. This implies that, for a sequence of $N$-point DCT blocks $\{X_i^{(N)}\}, i = 0, 1, \ldots, M-1$, there is a matrix $A_{(M,N)}$ of size $MN \times MN$ such that the corresponding composite DCT $X^{(MN)}$ ($MN$-point DCT) holds the following relation with $M$ block DCTs:

$$X^{(MN)} = A_{(M,N)}[X_0^{(N)} X_1^{(N)} \ldots X_{M-1}^{(N)}]^T. \qquad (2.109)$$

The matrix $A_{(M,N)}$ is given by

$$A_{(M,N)} = C_{MN} \begin{bmatrix} C_N^{-1} & 0_N & 0_N & \cdots & 0_N & 0_N \\ 0_N & C_N^{-1} & 0_N & \cdots & 0_N & 0_N \\ \vdots & \vdots & \vdots & \ddots & \vdots & \vdots \\ 0_N & 0_N & 0_N & \cdots & C_N^{-1} & 0_N \\ 0_N & 0_N & 0_N & \cdots & 0_N & C_N^{-1} \end{bmatrix}, \qquad (2.110)$$

where $0_N$ represents the $N \times N$ null matrix. A typical example of a conversion matrix is given below:

$$A_{(2,4)} = C_8 \begin{bmatrix} C_4^{-1} & 0_4 \\ 0_4 & C_4^{-1} \end{bmatrix}$$

$$= \begin{bmatrix} 0.7071 & 0 & 0 & 0 & 0.7071 & 0 & 0 & 0 \\ 0.6407 & 0.294 & -0.0528 & 0.0162 & -0.6407 & 0.294 & 0.0528 & 0.0162 \\ 0 & 0.7071 & 0 & 0 & 0 & 0.7071 & 0 & 0 \\ -0.225 & 0.5594 & 0.3629 & -0.0690 & 0.225 & 0.5594 & -0.3629 & -0.069 \\ 0 & 0 & 0.7071 & 0 & 0 & 0 & 0.7071 & 0 \\ 0.1503 & -0.2492 & 0.5432 & 0.3468 & -0.1503 & -0.2492 & -0.5432 & 0.3468 \\ 0 & 0 & 0 & 0.7071 & 0 & 0 & 0 & -0.7071 \\ -0.1274 & 0.1964 & -0.2654 & 0.6122 & 0.1274 & 0.1964 & 0.2654 & 0.6122 \end{bmatrix}.$$
$$(2.111)$$

We find that the conversion matrices and their inverses are sparse. Hence, this property is exploited to design a computational technique with fewer number of multiplications and additions of two numbers than those needed in the usual matrix multiplications. This is elaborated in the next subsection.

### 2.2.4.6   Properties of Block Composition Matrices

Let us partition a composition matrix $A_{(M,N)}$ into $M \times M$ submatrices, each of which has the dimension $N \times N$. We refer the $(i, j)th, 0 \leq i, j \leq M-1$ submatrix of $A_{(M,N)}$ as $A_{i,j}$. Similarly, the DCT matrix $C_{MN}$ is also partitioned into a set of $M \times M$ submatrices whose $(i, j)th, 0 \leq i, j \leq M-1$ submatrix is denoted by $(C_{MN})_{i,j}$. From Eq. (2.110),

$$A_{i,j} = (C_{MN})_{i,j} C_N^T. \qquad (2.112)$$

Two useful properties of block composition matrices are presented in the following theorems.

**Theorem 2.14** *Given odd $M$ and even $N$, elements of $A_{j, \frac{M-1}{2}}$ satisfy the following condition:*

$$A_{j, \frac{M-1}{2}}(k, l) = 0, \quad \text{if } k + l \text{ is odd.} \qquad (2.113)$$

□

**Proof:**

From Eq. (2.112), $A_{j,\frac{M-1}{2}} = (C_{MN})_{j,\frac{M-1}{2}} C_N^T$.

For the sake of convenience, let us rename $(C_{MN})_{j,\frac{M-1}{2}}$ as $S$. From Eq. (2.10), $S(k,l) = (-1)^k S(k, N-1-l)$.

Hence,

$$
\begin{aligned}
A_{j,\frac{M-1}{2}}(k,l) &= \sum_{t=0}^{N-1} S(k,t) C_N^T(t,l), \\
&= \sum_{t=0}^{\frac{N}{2}-1} (S(k,t) C_N(l,t) \\
&\qquad + (-1)^{(k+l)} S(k,t) C_N(l,t)), \\
&= \begin{cases} 2 \displaystyle\sum_{t=0}^{\frac{N}{2}-1} S(k,t) C_N(l,t), & k+l \text{ even}, \\ 0, & k+l \text{ odd}. \end{cases}
\end{aligned}
\tag{2.114}
$$

□

**Theorem 2.15** *Given even $N$, $A_{i,j}$ and $A_{i,M-1-j}$ are related as follows:*

$$
A_{i,j}(k,l) = (-1)^{k+l} A_{i,M-1-j}(k,l). \tag{2.115}
$$

□

**Proof:**

For convenience let us rename $(C_{MN})_{i,j}$ and $(C_{MN})_{i,M-1-j}$ as $S_0$ and $S_1$, respectively. From Eq. (2.10), $S_0(k,l) = (-1)^k S_1(k, N-1-l)$. Now,

$$
\begin{aligned}
A_{i,j}(k,l) &= \sum_{t=0}^{N-1} S_0(k,t) C_N(l,t), \\
&= \sum_{t=0}^{N-1} ((-1)^k S_1(k, N-1-t) \\
&\qquad (-1)^l C_N(l, N-1-t)), \\
&= \sum_{u=0}^{N-1} (-1)^{k+l} S_1(k,u) C_N(l,u), \\
&\qquad\qquad \text{Substituting } (N-1-t) \text{ as } u, \\
&= (-1)^{k+l} A_{i,M-1-j}(k,l).
\end{aligned}
\tag{2.116}
$$

□

In [67], it has been shown that $A_{(M,N)}$ is a sparse matrix where for every $M \times N$ blocks there are $N-1$ zeroes, and in total there are $MN(N-1)$ zeroes. The corresponding property is stated below.

**Theorem 2.16** *In* $A_{(M,N)}$ *every ith row such that,* $i \bmod M = 0$, $0 \le i \le MN - 1$, *contains* $MN - M$ *zeroes.*  □

From Theorems 2.16 and 2.14 we can state the following theorem about the number of zeroes in each row of the composition matrix when $M$ is odd.

**Theorem 2.17** *If* $M$ *is odd, the number of zeroes* $(n_z(i))$ *in the ith row of* $A_{(M,N)}$ *is given by the following expression.*

$$
\begin{aligned}
n_z(i) &= MN - M, & \text{if } i \bmod M = 0, \\
&= \tfrac{N}{2}, & \text{Otherwise.}
\end{aligned} \tag{2.117}
$$

□

Hence, the extent of sparseness of a composition matrix is stated in the following form.

**Theorem 2.18** *The total number of zeroes* $(n_z)$ *in* $A_{(M,N)}$ *is as follows:*

$$
\begin{aligned}
n_z &= MN(N - 1), & M \text{ even}, \\
&= \tfrac{N}{2}(3MN - 2M - N), & M \text{ odd}.
\end{aligned} \tag{2.118}
$$

□

**Proof:**

**M even:** There are $\frac{MN}{M} = N$ rows, which have $MN - M$ zeroes each. Hence, the total number of zeroes is $N(MN - M) = MN(N - 1)$.

**M odd:** In addition to the rows with $MN - M$ zeroes, there are $MN - N$ rows each having $\frac{N}{2}$ zeroes. Hence, the total number of zeroes is $N(MN - M) + (MN - N)\frac{N}{2} = \frac{N}{2}(3MN - 2M - N)$.

□

From the above properties we may have the number of operations required for performing block composition and decomposition operations. These are stated below.

**Theorem 2.19** *The number of multiplications* $(n_m)$ *and additions* $(n_a)$ *for composition of* $M$ *adjacent* $N$-*point DCT blocks are given below.*

$$
n_m = \frac{M(M - 1)}{2} N^2 + N. \tag{2.119}
$$

$$
\begin{aligned}
n_a &= (MN - N)MN, & M \text{ even}, \\
&= (MN - N)(MN - \tfrac{N}{2} - 1), & M \text{ odd}.
\end{aligned} \tag{2.120}
$$

□

**Proof:**

**M even:** As magnitudes of elements are repeated according to Theorem 2.15, for $MN - N$ rows it requires $\frac{MN}{2}$ numbers of multiplications. However, for them, the number of additions remains at $MN - 1$. Again, from Theorem 2.16 for $N$ rows, there are $M$ nonzero elements whose magnitudes are the same. Hence, each of them requires a multiplication and $M - 1$ additions. The numbers of multiplications and additions for even $M$ are as follows:

$$
\begin{aligned}
n_m &= (MN - N)\frac{MN}{2} + N, \\
&= \frac{M(M-1)}{2}N^2 + N.
\end{aligned}
\tag{2.121}
$$

$$
\begin{aligned}
n_a &= (MN - N)(MN - 1) + N(M - 1), \\
&= (MN - N)MN.
\end{aligned}
\tag{2.122}
$$

**M odd:** As magnitudes of elements are repeated according to Theorem 2.15, for $MN - N$ rows it requires $\frac{(M-1)}{2}N + \frac{N}{2} = \frac{MN}{2}$ numbers of multiplications. However, for them, the number of additions remains at $MN - \frac{N}{2} - 1$. Again, from Theorem 2.16 for $N$ rows, there are $M$ nonzero elements whose magnitudes are the same. Hence, each of them computes with 1 multiplication and $M - 1$ additions. This makes the numbers of multiplications and additions as

$$
\begin{aligned}
n_m &= (MN - N)\frac{MN}{2} + N, \\
&= \frac{M(M-1)}{2}N^2 + N.
\end{aligned}
\tag{2.123}
$$

$$
\begin{aligned}
n_a &= (MN - N)(MN - \frac{N}{2} - 1) + N(M - 1), \\
&= (MN - N)(MN - \frac{N}{2}).
\end{aligned}
\tag{2.124}
$$

$\square$

**Theorem 2.20** *The number of multiplications ($n_m$) and additions ($n_a$) for decomposition of an $MN$-point DCT block into $M$ adjacent $N$-point DCT blocks are as follows:*

$$
\begin{aligned}
n_m &= \frac{MN}{2}(MN - N + 1), & M \text{ even}, \\
&= \frac{(M-1)N}{2}(MN - N + 1) + \frac{M}{2}N^2, & M \text{ odd}.
\end{aligned}
\tag{2.125}
$$

$$
\begin{aligned}
n_a &= \frac{(MN)^2}{2}, & M \text{ even}, \\
&= \frac{M(M-1)}{2}N^2 + (\frac{MN}{2} - 1)N, & M \text{ odd}.
\end{aligned}
\tag{2.126}
$$

$\square$

**Proof:**

**M even:** In $A_{(M,N)}{}^T$, for every pair of $i$th and $i + \frac{M}{2}N$ rows, the magnitudes of corresponding elements of the same columns are the same (see Theorem 2.15). Exploiting this property, we have the following numbers of multiplications and additions for each pair. Moreover, due to Theorem 2.16, each row contains $N - 1$ zeroes.

$$
\begin{aligned}
n_m &= \frac{MN}{2}(MN - N + 1), \\
n_a &= \frac{MN}{2}.MN, \\
&= \frac{(MN)^2}{2}.
\end{aligned}
\tag{2.127}
$$

**M odd:** In this case, $N$ rows around the center of the transformation matrix have $\frac{MN}{2}$ zeroes. For other rows, they occur in pairs with the similar property of even $M$.

$$
\begin{aligned}
n_m &= \frac{(M-1)N}{2}(MN - N + 1) + \frac{MN}{2}N, \\
&= \frac{(M-1)N}{2}(MN - N + 1) + \frac{M}{2}N^2. \\
n_a &= \frac{(M-1)N}{2}MN + (\frac{MN}{2} - 1)N, \\
&= \frac{M(M-1)}{2}N^2 + (\frac{MN}{2} - 1)N.
\end{aligned}
\tag{2.128}
$$

□

### 2.2.4.7  Matrix Factorization

The $N$-point DCT matrix, $C_N$, could be factorized into a product of a few sparse matrices. This is given below:

$$
C_N = \frac{1}{\sqrt{2}}P_N
\begin{bmatrix}
C_{\frac{N}{2}} & 0_{\frac{N}{2}} \\
0_{\frac{N}{2}} & J_{\frac{N}{2}}C_{\frac{N}{2}}^{IV}J_{\frac{N}{2}}
\end{bmatrix}
\begin{bmatrix}
I_{\frac{N}{2}} & J_{\frac{N}{2}} \\
J_{\frac{N}{2}} & -I_{\frac{N}{2}}
\end{bmatrix},
\tag{2.129}
$$

where $P_N$ is the $N \times N$ permutation matrix as defined below:

$$
P_N =
\begin{bmatrix}
1 & 0 & 0 & \cdots & 0 & 0 & 0 & \cdots & 0 \\
0 & 0 & 0 & \cdots & 0 & 1 & 0 & \cdots & 0 \\
0 & 1 & 0 & \cdots & 0 & 0 & 0 & \cdots & 0 \\
0 & 0 & 0 & \cdots & 0 & 0 & 1 & \cdots & 0 \\
 & & & \cdots & & & & \cdots & \\
0 & 0 & 0 & \cdots & 1 & 0 & 0 & \cdots & 0 \\
0 & 0 & 0 & \cdots & 0 & 0 & 0 & \cdots & 1
\end{bmatrix}.
\tag{2.130}
$$

$I_N$, $J_N$ and $0_N$ are $N \times N$ *identity*, *reverse identity*, and *zero* matrices, respectively.

### 2.2.4.8  8-Point Type-II DCT Matrix ($C_8$)

As the $8 \times 8$ block DCT is used in JPEG and MPEG compression techniques, a few specific properties of its transform matrix ($C_8$) are worth reviewing.

• **Factorization:** $C_8$ may be factorized as a product of matrices [41], that are mostly sparse, and some of them contain either 1 or $-1$ as nonzero elements. These properties are useful in designing fast matrix multiplication operations for various algorithms.

$$C_8 = DPB_1B_2MA_1A_2A_3, \qquad (2.131)$$

where, $D = \mathbb{D}\left([0.3536\ 0.2549\ 0.2706\ 0.3007\ 0.3536\ 0.4500\ 0.6533\ 1.2814]^T\right)$. $\mathbb{D}(\mathbf{x})$ denotes a *diagonal matrix* whose elements of its diagonal are formed by the column vector $\mathbf{x}$. Definitions of other matrices in Eq. (2.131) are given below.

$$P = \begin{bmatrix} 1 & 0 & 0 & 0 & 0 & 0 & 0 & 0 \\ 0 & 0 & 0 & 0 & 0 & 1 & 0 & 0 \\ 0 & 0 & 1 & 0 & 0 & 0 & 0 & 0 \\ 0 & 0 & 0 & 0 & 0 & 0 & 0 & 1 \\ 0 & 1 & 0 & 0 & 0 & 0 & 0 & 0 \\ 0 & 0 & 0 & 0 & 1 & 0 & 0 & 0 \\ 0 & 0 & 0 & 1 & 0 & 0 & 0 & 0 \\ 0 & 0 & 0 & 0 & 0 & 0 & 1 & 0 \end{bmatrix},$$

$$B1 = \begin{bmatrix} 1 & 0 & 0 & 0 & 0 & 0 & 0 & 0 \\ 0 & 1 & 0 & 0 & 0 & 0 & 0 & 0 \\ 0 & 0 & 1 & 0 & 0 & 0 & 0 & 0 \\ 0 & 0 & 0 & 1 & 0 & 0 & 0 & 0 \\ 0 & 0 & 0 & 0 & 1 & 0 & 0 & 1 \\ 0 & 0 & 0 & 0 & 0 & 1 & 1 & 0 \\ 0 & 0 & 0 & 0 & 1 & -1 & 0 \\ 0 & 0 & 0 & 0 & -1 & 0 & 0 & 1 \end{bmatrix}, B2 = \begin{bmatrix} 1 & 0 & 0 & 0 & 0 & 0 & 0 & 0 \\ 0 & 1 & 0 & 0 & 0 & 0 & 0 & 0 \\ 0 & 0 & 1 & 1 & 0 & 0 & 0 & 0 \\ 0 & 0 & -1 & 1 & 0 & 0 & 0 & 0 \\ 0 & 0 & 0 & 0 & 1 & 0 & 0 & 0 \\ 0 & 0 & 0 & 0 & 0 & 1 & 0 & 1 \\ 0 & 0 & 0 & 0 & 0 & 0 & 1 & 0 \\ 0 & 0 & 0 & 0 & 0 & -1 & 0 & 1 \end{bmatrix},$$

$$M = \begin{bmatrix} 1 & 0 & 0 & 0 & 0 & 0 & 0 & 0 \\ 0 & 1 & 0 & 0 & 0 & 0 & 0 & 0 \\ 0 & 0 & 0.7071 & 0 & 0 & 0 & 0 & 0 \\ 0 & 0 & 0 & 1 & 0 & 0 & 0 & 0 \\ 0 & 0 & 0 & 0 & -0.9239 & 0 & -0.3827 & 0 \\ 0 & 0 & 0 & 0 & 0 & 0.7071 & 0 & 0 \\ 0 & 0 & 0 & 0 & -0.3827 & 0 & 0.9239 & 0 \\ 0 & 0 & 0 & 0 & 0 & 0 & 0 & 0 \end{bmatrix},$$

$$A_1 = \begin{bmatrix} 1 & 1 & 0 & 0 & 0 & 0 & 0 & 0 \\ 1 & -1 & 0 & 0 & 0 & 0 & 0 & 0 \\ 0 & 0 & 1 & 1 & 0 & 0 & 0 & 0 \\ 0 & 0 & 0 & 1 & 0 & 0 & 0 & 0 \\ 0 & 0 & 0 & 0 & 1 & 0 & 0 & 0 \\ 0 & 0 & 0 & 0 & 0 & 1 & 0 & 0 \\ 0 & 0 & 0 & 0 & 0 & 0 & 1 & 0 \\ 0 & 0 & 0 & 0 & 0 & 0 & 0 & 1 \end{bmatrix},$$

$$A_2 = \begin{bmatrix} 1 & 0 & 0 & 1 & 0 & 0 & 0 & 0 \\ 0 & 1 & 1 & 0 & 0 & 0 & 0 & 0 \\ 0 & 1 & -1 & 0 & 0 & 0 & 0 & 0 \\ 1 & 0 & 0 & -1 & 0 & 0 & 0 & 0 \\ 0 & 0 & 0 & 0 & -1 & -1 & 0 & 0 \\ 0 & 0 & 0 & 0 & 0 & 1 & 1 & 0 \\ 0 & 0 & 0 & 0 & 0 & 0 & 1 & 1 \\ 0 & 0 & 0 & 0 & 0 & 0 & 0 & 1 \end{bmatrix}, \text{ and}$$

$$A_3 = \begin{bmatrix} 1 & 0 & 0 & 0 & 0 & 0 & 0 & 1 \\ 0 & 1 & 0 & 0 & 0 & 0 & 1 & 0 \\ 0 & 0 & 1 & 0 & 0 & 1 & 0 & 0 \\ 0 & 0 & 0 & 1 & 1 & 0 & 0 & 0 \\ 0 & 0 & 0 & 1 & -1 & 0 & 0 & 0 \\ 0 & 0 & 1 & 0 & 0 & -1 & 0 & 0 \\ 0 & 1 & 0 & 0 & 0 & 0 & -1 & 0 \\ 1 & 0 & 0 & 0 & 0 & 0 & 0 & -1 \end{bmatrix}.$$

#### 2.2.4.9 Integer Cosine Transforms

*Integer cosine transforms* (ICT) [18] are derived from a DCT matrix as its integer approximations and preserving its properties of orthogonality, symmetry, relative order, and sign of the elements of the matrix. For example, from the $C_8$ DCT matrix we may obtain the following general form of a matrix preserving the symmetry and signs of its elements.

$$T_8(a,b,c,d,e,f) = \begin{bmatrix} 1 & 1 & 1 & 1 & 1 & 1 & 1 & 1 \\ a & b & c & d & -d & -c & -b & -a \\ e & f & -f & -e & -e & -f & f & e \\ b & -d & -a & -c & c & a & d & -b \\ 1 & -1 & -1 & 1 & 1 & -1 & -1 & 1 \\ c & -a & d & b & -b & -d & a & -c \\ f & -e & e & -f & -f & e & -e & f \\ d & -c & b & -a & a & -b & c & -d \end{bmatrix}, \quad (2.132)$$

where $a$, $b$, $c$, $d$, $e$, and $f$ are integers satisfying the following constraints for preserving the orthogonal property and relative order of their magnitude corresponding to elements in $C_8$ at the same locations of the matrix.

$$\begin{aligned} ab &= ac + bd + cd, \\ a \geq b \geq c \geq d \quad and \quad e &\geq f. \end{aligned} \quad (2.133)$$

Hence, an infinite number of transform matrices could be derived satisfying the above constraints. A typical example of one such matrix with $a = 5$, $b = 3$, $c = 2$, $d = 1$, $e = 3$, and $f = 1$ are given below. Magnitudes of all the elements in the following transform is in the form of $2^n$ or $2^n + 1$ where $n$ is an integer. This makes multiplication with these elements quite fast, as it can be implemented using *shift and add* method.

$$T_8(5,3,2,1,3,1) = \begin{bmatrix} 1 & 1 & 1 & 1 & 1 & 1 & 1 & 1 \\ 5 & 3 & 2 & 1 & -1 & -2 & -3 & -5 \\ 3 & 1 & -1 & -3 & -3 & -1 & 1 & 3 \\ 3 & -1 & -5 & -2 & 2 & 5 & 1 & -3 \\ 1 & -1 & -1 & 1 & 1 & -1 & -1 & 1 \\ 2 & -5 & 1 & 3 & -3 & -1 & 5 & -2 \\ 1 & -3 & 3 & -1 & -1 & 3 & -3 & 1 \\ 1 & -2 & 3 & -5 & 5 & -3 & 2 & -1 \end{bmatrix}. \quad (2.134)$$

For making the transform orthonormal, we have to divide each row vector by normalizing scalar constants. This means that, after the transform, the transform coefficients should be divided by the corresponding magnitudes of each row vector of the transform matrix. This, in fact, is taken care of during the quantization process so that it does not introduce any additional overhead in the computation of quantized coefficients for compression of data. For example, the normalizing constants for the row vectors of $T_8(5,3,2,1,3,1)$ are 8, 78, 40, 78, 8, 78, 40, and 78, respectively.

Following similar approaches, different integer cosine matrices for other DCT matrices may be obtained. Out of them, in the H.264, the following ICT matrix $(T_4)$ corresponding to the 4-point DCT matrix $(C_4)$ is used.

$$T_4 = \begin{bmatrix} 1 & 1 & 1 & 1 \\ 2 & 1 & -1 & -2 \\ 1 & -1 & -1 & 1 \\ 1 & -2 & 2 & -1 \end{bmatrix}. \tag{2.135}$$

The inverse of $T_4$ also has elements suitable for integer arithmetics during the matrix multiplication operation.

## 2.2.5 Hadamard Transform

In the Hadamard transform [52] of a sequence of length $2^m, m \in \mathbb{Z}$, elements of the transform matrix $(Hd_m)$ are either 1 or $-1$. The matrix is uniformly scaled so that it becomes orthonormal. The Hadamard transform matrix can be defined recursively in the following way:

$$\begin{aligned} Hd_0 &= 1, \\ Hd_m &= \frac{1}{\sqrt{2}} \begin{bmatrix} Hd_{m-1} & Hd_{m-1} \\ Hd_{m-1} & -Hd_{m-1} \end{bmatrix}. \end{aligned} \tag{2.136}$$

The advantage of the Hadamard transform is that it is a real transform and does not require multiplication operation in its computation.

## 2.2.6 Discrete Wavelet Transform (DWT)

Previously we have discussed that there exist two mother wavelets $\phi(x)$ and $\psi(x)$ such that their translated and dilated forms provide an orthonormal basis. In this section we discuss the properties of these two functions defined over a finite discretized grid $\mathbb{Z}_N = \{0, 1, 2, \ldots, N-1\}$.

### 2.2.6.1 Orthonormal Basis with a Single Mother Wavelet

From a mother wavelet $\phi(n)$, the family of translated basis vectors is defined as $\{\phi(n-k)|k \in \mathbb{Z}_N\}$. The following theorem states when this family forms the orthonormal basis in $\mathbb{Z}_N$ [45].

**Theorem 2.21** *The orthonormal basis for a finite length sequence of length $N$ is formed from a mother wavelet $\phi(n)$ by its translation at every point in $\mathbb{Z}_N$ if and only if $|\hat{\phi}(n)| = 1$ for all $n \in \mathbb{Z}_N$, where $\hat{\phi}(n)$ is the DFT of $\phi(n)$.*

□

The above theorem implies that we would not be able to achieve frequency localization for a mother wavelet. In fact, as the magnitude of the spectrum becomes 1 for every frequency, the mother wavelet is the *Dirac delta function* or $\delta(n)$, which is the same trivial orthonormal expansion as discussed earlier. However, we may overcome this constraint when *two* mother wavelets (or sometimes denoted as *parent* wavelets) are used as discussed in the following.

## 2.2.6.2   Orthonormal Basis with Two Mother Wavelets

In this case, we consider sequences whose lengths are even numbers (say, $N = 2M$, $M, N \in \mathbb{Z}$). The following theorem states the conditions for existence of parent wavelets for forming an orthonormal basis [45] [93].

**Theorem 2.22** *For $N = 2M$, there exist two mother wavelets $\phi(n)$ and $\psi(n)$ such that the union of basis sets $B_1 = \{\phi(n - 2k)|k \in \mathbb{Z}_M\}$ and $B_2 = \{\phi(n - 2k)|k \in \mathbb{Z}_M\}$ with cardinality of each set being $M$ forms an orthonormal basis. The basis set will be orthonormal if and only if the following conditions are satisfied for all $n \in \mathbb{Z}_N$:*

$$
\begin{aligned}
|\hat{\phi}(n)|^2 + |\hat{\phi}(n + M)|^2 &= 2, \\
|\hat{\psi}(n)|^2 + |\hat{\psi}(n + M)|^2 &= 2, \\
\hat{\phi}(n)\hat{\psi}^*(n) + \hat{\phi}(n + M)\hat{\psi}^*(n + M) &= 0.
\end{aligned}
\tag{2.137}
$$

$\square$

Let us reflect on the nature of these two mother wavelets from the constraints imposed upon by Eq. (2.137). According to this equation, the average of $|\hat{\phi}(n)|^2$ and $|\hat{\phi}(n + M)|^2$ should be 1. Suppose $|\hat{\phi}(n)|^2$ is set to 2. In that case, $|\hat{\phi}(n+M)|^2$ should be zero (0). Hence, $\phi(n)$ behaves like a *low-pass filter* function. However, the third constraint of Eq. (2.137) enforces $|\hat{\psi}(n)|$ to be zero, thus implying $|\hat{\psi}(n + M)|^2 = 2$. Hence, $\psi(n)$ acts like a *high-pass filter*. The transform obtained from the orthonormal basis set $B_1$, which is defined from $\phi(n)$ (with low-pass filter response), provides *the approximation* of the function at *half* the resolution of the original one. The approximated function at lower resolution may further be transformed using the translated and dilated forms of mother wavelets $\phi(n)$ and $\psi(n)$ in the next level. The *high-pass filtering function* $\psi(n)$ provides the orthonormal basis ($B_2$) for transforming the function into its *details* at the given resolution. In a *dyadic expansion* of a function, at each stage only the *details* are retained leaving aside its *approximation* at the lowest level of resolution. Let us discuss the well-known Haar wavelets under this context.

## 2.2.6.3   Haar Wavelets

We illustrate the *discrete Haar wavelet transform* for $N = 8$. In $\mathbb{Z}_8$, the pair of Haar wavelets forming the orthonormal basis are as follows:

$$
\begin{aligned}
\phi(n) &= \tfrac{1}{\sqrt{2}}(1, 1, 0, 0, 0, 0, 0, 0), \\
\psi(n) &= \tfrac{1}{\sqrt{2}}(1, -1, 0, 0, 0, 0, 0, 0).
\end{aligned}
$$

Both these functions satisfy Eq. (2.137). So the transformation matrix ($W_8$) for the first stage of wavelet decomposition (into *approximation* and *details*) of a sequence $x(n)$ of length 8 is given by the following.

$$W_8 = \frac{1}{\sqrt{2}} \begin{bmatrix} 1 & 1 & 0 & 0 & 0 & 0 & 0 & 0 \\ 0 & 0 & 1 & 1 & 0 & 0 & 0 & 0 \\ 0 & 0 & 0 & 0 & 1 & 1 & 0 & 0 \\ 0 & 0 & 0 & 0 & 0 & 0 & 1 & 1 \\ 1 & -1 & 0 & 0 & 0 & 0 & 0 & 0 \\ 0 & 0 & 1 & -1 & 0 & 0 & 0 & 0 \\ 0 & 0 & 0 & 0 & 1 & -1 & 0 & 0 \\ 0 & 0 & 0 & 0 & 0 & 0 & 1 & -1 \end{bmatrix}. \tag{2.138}$$

The first four rows of $W_8$ correspond to the basis vectors formed from $\phi(n)$, whereas the last four rows are obtained from the $\psi(n)$ mother wavelet. The transform $(X = W_8 x)$ provides the first four coefficients as the *approximation coefficients*, and the last four as the *detail coefficients*. In the same way, in successive stages, approximation coefficients are further transformed by $W_4$ and $W_2$ formed in the same way as in Eq. (2.138). Hence, in a dyadic expansion, the complete Haar transform can be expressed in the following way:

$$X = \begin{bmatrix} W_2 & 0_{2\times 6} \\ 0_{6\times 2} & I_6 \end{bmatrix} \begin{bmatrix} W_4 & 0_4 \\ 0_4 & I_4 \end{bmatrix} W_8 x \tag{2.139}$$

In the above equation, $0_{m\times n}$ denotes the null matrix of dimension $m \times n$ where $m$ and $n$ are arbitrary integers. As previously, $I_n$ and $0_n$ denote the identity matrix and the null matrix, respectively, of dimension $n \times n$.

From Eq. (2.139) we derive the transform matrix for the discrete Haar transform (for $N = 8$) as follows:

$$\mathbb{H}_8 = \begin{bmatrix} \frac{1}{\sqrt{16}} & \frac{1}{\sqrt{16}} & \frac{1}{\sqrt{16}} & \frac{1}{\sqrt{16}} & \frac{1}{\sqrt{16}} & \frac{1}{\sqrt{16}} & \frac{1}{\sqrt{16}} & \frac{1}{\sqrt{16}} \\ \frac{1}{\sqrt{8}} & \frac{1}{\sqrt{8}} & \frac{1}{\sqrt{8}} & \frac{1}{\sqrt{8}} & -\frac{1}{\sqrt{8}} & -\frac{1}{\sqrt{8}} & -\frac{1}{\sqrt{8}} & -\frac{1}{\sqrt{8}} \\ \frac{1}{\sqrt{4}} & \frac{1}{\sqrt{4}} & -\frac{1}{\sqrt{4}} & -\frac{1}{\sqrt{4}} & 0 & 0 & 0 & 0 \\ 0 & 0 & 0 & 0 & \frac{1}{\sqrt{4}} & \frac{1}{\sqrt{4}} & -\frac{1}{\sqrt{4}} & -\frac{1}{\sqrt{4}} \\ \frac{1}{\sqrt{2}} & -\frac{1}{\sqrt{2}} & 0 & 0 & 0 & 0 & 0 & 0 \\ 0 & 0 & \frac{1}{\sqrt{2}} & -\frac{1}{\sqrt{2}} & 0 & 0 & 0 & 0 \\ 0 & 0 & 0 & 0 & \frac{1}{\sqrt{2}} & -\frac{1}{\sqrt{2}} & 0 & 0 \\ 0 & 0 & 0 & 0 & 0 & 0 & \frac{1}{\sqrt{2}} & -\frac{1}{\sqrt{2}} \end{bmatrix}. \tag{2.140}$$

The last seven basis vectors in $\mathbb{H}_8$ are formed from the dilated and translated form of $\psi(n)$ (that is, $2^{-\frac{j}{2}}\psi(2^{-j}n - 2k), j \in \{1, 2, 3\}, k = 1, 2, .., \frac{N}{2^j}$. The first row corresponds to the scaling function and computes the average of the function. We may extend this observation to construct the *Haar transform matrix* for other values of $N = 2^m, m \in \mathbb{Z}$.

#### 2.2.6.4  Other Wavelets

Like the Haar wavelets, other wavelet functions may be obtained by satisfying Eq. (2.137). To make these functions real, the frequency transforms for $\phi(n)$

and $\psi(n)$ should be chosen such that $\widehat{\phi}(n) = \widehat{\phi}^*(N-n)$ and $\widehat{\psi}(n) = \widehat{\psi}^*(N-n)$. In fact, given an orthonormal mother wavelet $\phi(n)$, there exists a wavelet function $\psi(n)$ satisfying the other two conditions of Eq. (2.137) for the orthonormal basis. The following theorem states this relationship.

**Theorem 2.23** *Suppose $N = 2M$ (N and M are positive integers), and $\phi(n)$ is a function such that $\{\phi(n-2k)|k = 0, 1, \ldots, M\}$ forms an orthonormal basis. Let us define the function $\psi(n)$ as follows:*

$$\psi(n) = (-1)^{n-1}\phi^*(1-n).$$

*In that case, $\{\phi(n-2k)|k = 0, 1, \ldots, M\}$ and $\{\psi(n-2k)|k = 0, 1, \ldots, M\}$ generate the orthonormal basis in $\mathbb{Z}_N$.* □

The above theorem implies that orthonormal wavelet basis vectors can be specified by a single mother wavelet function. A few more examples of such scaling functions $(\phi(n))$ that have a compact support with real values are given below. These are from the family of Daubechies' orthogonal wavelets [31], which can be formed for any even positive integer $p$, where the nonzero elements in the function occur only for $0 \leq n < p$. Two such functions for $p = 4$ and $6$ are given below.

$$
\begin{aligned}
D_4(n) &= \tfrac{\sqrt{2}}{8}(1+\sqrt{3}, 3+\sqrt{3}, 3-\sqrt{3}, 1-\sqrt{3}), \\
D_6(n) &= \tfrac{\sqrt{2}}{32}(b+c, 2a+3b+3c, 6a+4b+2c, 6a+4b-2c, 2a+3b-3c, b-c).
\end{aligned}
$$
$$(2.141)$$

In the above equation, $a = 1 - \sqrt{10}, b = 1 + \sqrt{10}$ and $c = \sqrt{5 + 2\sqrt{10}}$.

### 2.2.6.5 DWT through Filter Banks

The DWT can be computed efficiently using filter banks, a *low-pass filter* for computing *the approximation coefficients*, and a *high-pass filter* for computing the *detailed coefficients*. We can show that computation of *inner products* with the orthonormal basis vectors as obtained from two mother wavelets $\phi(n)$ and $\psi(n)$, and is equivalent to the convolution operation (or filtering) with a pair of filters (say, with impulse responses $h(n)$ and $g(n)$, respectively). Let us define *the conjugate reflection* of a function $w(n)$ as

**Definition 2.9** *For any $w(n) \in \mathbb{Z}_N$, its* conjugate reflection $\overline{w}(n)$ *is defined as*

$$\overline{w}(n) = w^*(-n) = w^*(N-n) \text{ for all } n \in \mathbb{Z}_N.$$

□

**Theorem 2.24** *Let $x(n)$ be a sequence of length $N$ (or $x(n) \in \mathbb{Z}_N$), and $w(n)$ be a function in $\mathbb{Z}_N$. Then,*

$$
\begin{aligned}
x \circledast \overline{w}(k) &= <x(n), w(n-k)>, \\
x \circledast w(k) &= <x(n), \overline{w}(n-k)>,
\end{aligned}
$$
$$(2.142)$$

*where $< a(n), b(n) >$ denotes the dot product between $a(n)$ and $b(n)$ in $\mathbb{Z}_N$.* $\square$

From the foregoing theorem the wavelet coefficients using the orthonormal basis from $\phi(n)$ and $\psi(n)$ are computed by using their *conjugate mirror filters* [93] as follows.

**Theorem 2.25** *Suppose $h(n) = \overline{\phi}(n)$ and $g(n) = \overline{\psi}(n)$ are two corresponding mother wavelets. Then approximation ($a_k, 0 \leq k < M$) and detail ($d_k, 0 \leq k < M$) coefficients of a function $x(n) \in \mathbb{Z}_N$ are obtained as follows:*

$$
\begin{aligned}
a_k &= x \circledast h(2k) &= &< x(n), \phi(n-2k) >,\\
d_k &= x \circledast g(2k) &= &< x(n), \psi(n-2k) > .
\end{aligned}
\tag{2.143}
$$

*This implies that downsampled $x \circledast h(n)$ and $x \circledast g(n)$ with a downsampling factor 2 (by retaining even samples and dropping odd samples in a sequence) provide the first-level decomposition of $x(n)$ into wavelet coefficients.* $\square$

From Theorem 2.23 we find that $g(n) = (-1)^{n-1} h^*(1-n)$. Reconstruction of $x(n)$ from wavelet coefficients is also possible by using a pair of filters, as shown in Figure 2.5. The process of transformation of $x(n)$ into a set of wavelet coefficients is also known as *analysis* of the function, whereas the inverse operation is known as *synthesis*. In the analysis stage, the filtered sequence is downsampled by a factor of 2, whereas in synthesis, input sequences are upsampled by the same factor before filtering. There exist combinations of $h(n)$, $g(n)$, $h'(n)$ and $g'(n)$ for which perfect reconstruction (that is, $x_0(n) = \widetilde{x_0}(n)$) is possible. This is not only true for orthonormal transforms, but also for non-orthogonal transforms (also known as *biorthonormal wavelet transforms* in this specific case). We consider here the relations among these filter banks for perfect reconstruction.

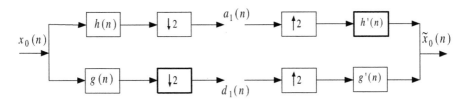

**Figure 2.5**: The discrete wavelet transform and its inverse using filter banks.(Reproduced from Figure 1.10 of Chapter 1.)

**Definition 2.10** *The even downsampling of $x(n)$ by a factor of 2 is defined as retaining of the even-indexed samples and dropping the odd ones. The operation is defined below.*

$$
D_e(x)(n) = x(2n).
\tag{2.144}
$$

*In the same way, the* odd downsampling *of $x(n)$ by a factor of 2 is defined as retaining of the odd-indexed samples and dropping the even ones. Mathematically, the operation is defined below.*

$$D_o(x)(n) = x(2n+1). \tag{2.145}$$

□

**Definition 2.11** *The upsampling of $x(n)$ by a factor of 2 is defined as inserting zeroes either between an* odd *and* even *sample, or between an* even *and* odd *sample. The first one we refer to as* even upsampling *(denoted by $U_e(.)$), and the latter as* odd upsampling *(denoted by $U_o(.)$). These operations are defined as follows.*

$$U_e(x)(m) = \begin{cases} x(n), & \text{if } m = 2n, \\ 0, & \text{otherwise.} \end{cases} \tag{2.146}$$

$$U_o(x)(m) = \begin{cases} x(n), & \text{if } m = 2n+1, \\ 0, & \text{otherwise.} \end{cases} \tag{2.147}$$

*If it is not mentioned, we consider the* upsampling *operation as even upsampling, and the operation is denoted by $U(.)$.*
□

According to Figure 2.5, perfect reconstruction of the input sequence implies that it should satisfy the following.

$$\tilde{x}_0(n) = h' \circledast U_e(D_e(h \circledast x_0))(n) + g' \circledast U_o(D_o(g \circledast x_0))(n) \tag{2.148}$$

For the above, the filter banks have the following relationship.

$$\begin{aligned} \widehat{h'}(n)\widehat{h}(n) + \widehat{g'}(n)\widehat{g}(n) &= 2, \\ \widehat{h'}(n)\widehat{h}(n+M) + \widehat{g'}(n)\widehat{g}(n+M) &= 0. \end{aligned} \tag{2.149}$$

The above relationship can be further simplified with orthonormal wavelet expansion as shown below.

$$\begin{aligned} h'(n) &= \overline{h}(n), \\ g'(n) &= \overline{g}(n). \end{aligned} \tag{2.150}$$

In a more general form accommodating the delay and gain in the final reconstruction of the function, the conditions in Eq. (2.149) can be expressed using the z-Transform as follows [100]:

$$\begin{aligned} H(z)H'(z) + G(z)G'(z) &= 2.c.z^{-l}, \\ H'(z)H(-z) + G'(z)G(-z) &= 0. \end{aligned} \tag{2.151}$$

where $H(z)$, $G(z)$, $H'(z)$, and $G(z)$ are the z-Transforms of $h(n)$, $g(n)$, $h'(n)$, and $g'(n)$, respectively. $c$ is a positive real constant, and $l$ is a nonnegative

**Table 2.7**: Daubechies 9/7 analysis and synthesis filter banks for lossy compression

| $n$ | Analysis filter bank | | Synthesis filter bank | |
|---|---|---|---|---|
| | $h(n)$ | $g(n-1)$ | $h'(n)$ | $g'(n+1)$ |
| 0 | 0.603 | 1.115 | 1.115 | 0.603 |
| $\pm 1$ | 0.267 | -0.591 | 0.591 | -0.267 |
| $\pm 2$ | -0.078 | -0.058 | -0.058 | -0.078 |
| $\pm 3$ | -0.017 | 0.091 | -0.091 | 0.017 |
| $\pm 4$ | 0.027 | | | 0.027 |

Table 2.8: 5/3 analysis and synthesis filter banks for lossless compression

| $n$ | Analysis filter bank | | Synthesis filter bank | |
|---|---|---|---|---|
| | $h(n)$ | $g(n-1)$ | $h'(n)$ | $g'(n+1)$ |
| 0 | $\frac{6}{8}$ | 1 | 1 | $\frac{6}{8}$ |
| $\pm 1$ | $\frac{2}{8}$ | $-\frac{1}{2}$ | $\frac{1}{2}$ | $-\frac{2}{8}$ |
| $\pm 2$ | $-\frac{1}{8}$ | | | $-\frac{1}{8}$ |

integer. The reconstructed output $(\tilde{x}_0(n))$, in that case, will become $cx_0(n-l)$. The second part of Eq. (2.151) is satisfied if the biorthogonal filters are chosen in the following way:

$$
\begin{aligned}
H'(z) &= -cz^{-m}G(-z), \\
G'(z) &= cz^{-m}H(-z).
\end{aligned} \tag{2.152}
$$

In JPEG2000 biorthogonal wavelets are used. Filter banks corresponding to these wavelets are shown in Tables 2.7 [4] and 2.8 [48]. We find that these filters satisfy Eq. (2.151) for $m = 1$ and $c = 1.0$. Because of this, in the analysis stage, the downsampled sequence from the low-pass filter $(h(n))$ retains only the even-indexed sample, whereas downsampling from the high-pass one $(g(n))$ keeps the odd-indexed sample [119]. Eventually, in the synthesis stage, corresponding upsampling becomes *even* and *odd* upsampling, respectively.

### 2.2.6.6 Lifting-based DWT

The lifting operation involves polyphase (or in our context, two-channel) decomposition of a sequence and prediction of its one channel from the other. This removes redundancy in that stream. In a dual lifting operation, subsequently the unprocessed part of the sequence is also modified or updated from the residual information of the other channel. For example, as shown in Figure 2.6, the sequence $x(n)$ is split into two phases or channels comprising sequences of *even* and *odd* samples $(x_e(n)$ and $x_o(n))$, respectively. In the first stage of lifting, odd samples are predicted from even samples, and residues are computed. In the next stage, those are used to update the even sequence $(x_e(n))$ so that the resulting sequences are in the form of $a(n)$ and $d(n)$. Perfect reconstruction is possible by performing the inverse operations as shown in Figure 2.7.

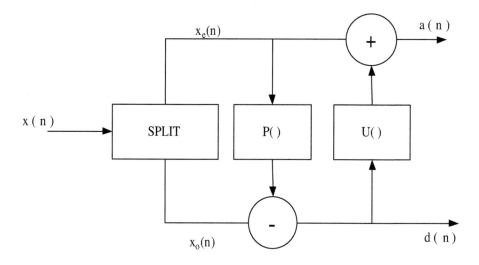

Figure 2.6: Dual lifting operations.

Wavelet filters can be implemented by a cascade of such basic lifting blocks. This makes the computation efficient. Various combinations of predictor $(P(.))$ and update $(U(.))$ functions implement these wavelet filters. For example, implementation of Daubechies 9/7 filters and Lee Gall's 5/3 integer wavelets are provided in Eq. (2.153) and (2.154), respectively.

**Lifting scheme for 9/7 filters**

$$
\begin{aligned}
a^{(0)}(l) &= x(2l), \\
d^{(0)}(l) &= x(2l+1), \\
d^{(1)}(l) &= d^{(0)}(l) + \alpha(a^{(0)}(l) + a^{(0)}(l+1)), \\
a^{(1)}(l) &= a^{(0)}(l) + \beta(d^{(1)}(l) + d^{(1)}(l-1)), \\
d^{(2)}(l) &= d^{(1)}(l) + \gamma(a^{(1)}(l) + a^{(1)}(l+1)), \\
a^{(2)}(l) &= a^{(1)}(l) + \delta(d^{(2)}(l) + d^{(2)}(l-1)), \\
a(l) &= \zeta a^{(2)}(l), \\
d(l) &= d^{(2)}(l)/\zeta,
\end{aligned}
\tag{2.153}
$$

where $\alpha \approx -1.59$, $\beta \approx -0.053$, $\gamma \approx 0.88$, $\delta \approx 0.44$, and $\zeta \approx 1.15$.

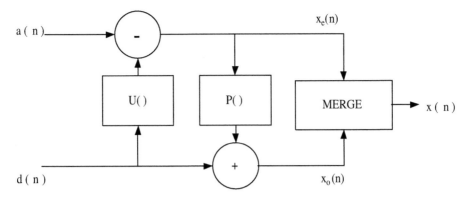

Figure 2.7: Inverse lifting operations.

**Lifting scheme for $5/3$ filters**

$$
\begin{aligned}
a^{(0)}(l) &= x(2l), \\
d^{(0)}(l) &= x(2l+1), \\
d(l) &= d^{(0)}(l) - \left\lfloor \frac{(a^{(0)}(l)+a^{(0)}(l+1))}{2} \right\rfloor, \text{ and }, \\
a(l) &= a^{(0)}(l) + \left\lfloor \frac{(d(l)+d(l-1))}{4} \right\rfloor.
\end{aligned} \tag{2.154}
$$

## 2.3   Transforms in 2-D Space

As images are functions in a 2-D space, 2-D transforms are more relevant to our discussion. In fact, all the above transforms in 1-D could be extended to 2-D, if the basis functions are *separable* in each dimension [52]. This implies that a function $b(x,y)$ can be expressed as a product of two 1-D functions of variables along each dimension (e.g., $b(x,y) = \phi(x)\psi(y)$).

This is also true for *discrete transforms*. A finite 2-D discrete function $f(m,n), 0 \le m < M, 0 \le n < N$ can also be represented by a *matrix* of dimension $M \times N$. Let $\Phi = \{\phi_i(m)|0 \le i, m < M\}$ and $\Psi = \{\psi_j(n)|0 \le j, n < N\}$ be two sets of orthonormal *discrete functions* in 1-D. It could be easily shown that $\Omega = \{\omega_{ij}(m,n) = \phi_i(m)\psi_j(n)|0 \le i, m < M, 0 \le j, n < N\}$ form *the orthonormal basis* of the 2-D discrete functions of dimension $M \times N$. Hence, the 2-D transform coefficients for expanding $f(m,n)$ with these basis functions could be computed as follows:

$$
\begin{aligned}
c(l,k) &= \sum_{m=0}^{M-1}\sum_{n=0}^{N-1} f(m,n)\omega_{ij}^{*}(m,n), \\
&= \sum_{m=0}^{M-1}\sum_{n=0}^{N-1} f(m,n)\phi_{i}^{*}(m)\psi_{j}^{*}(n), \\
&= \sum_{m=0}^{M-1} \phi_{i}^{*}(m) \left( \sum_{n=0}^{N-1} f(m,n)\psi_{j}^{*}(n) \right), \\
&\quad \text{for } 0 \le l \le M-1, \text{ and } 0 \le k \le N-1.
\end{aligned}
\tag{2.155}
$$

Similar to 1-D transform (see Eq. (2.44)), the foregoing equation can also be expressed in the form of a matrix transformation. Consider the matrix representation of the function $f(m,n)$ as $\mathbf{f}$ of dimension $M \times N$. The transform matrices of basis functions $\Phi$ and $\Psi$ are also represented in the following form.

$$
\Phi = \begin{bmatrix} \phi_{0}^{*\mathbf{T}} \\ \phi_{1}^{*\mathbf{T}} \\ \cdot \\ \cdot \\ \cdot \\ \phi_{M-1}^{*\mathbf{T}} \end{bmatrix}.
\tag{2.156}
$$

$$
\Psi = \begin{bmatrix} \psi_{0}^{*\mathbf{T}} \\ \psi_{1}^{*\mathbf{T}} \\ \cdot \\ \cdot \\ \cdot \\ \psi_{N-1}^{*\mathbf{T}} \end{bmatrix}.
\tag{2.157}
$$

Then the 2-D transform of $\mathbf{f}$ is given by Eq. (2.158).

$$
\mathbf{F} = \Phi \mathbf{f} \Psi^{\mathbf{T}}.
\tag{2.158}
$$

Given the property of orthonormal matrices, the inverse transform can be expressed as

$$
\mathbf{f} = \Phi^{\mathbf{H}} \mathbf{F} \left( \Psi^{\mathbf{H}} \right)^{\mathbf{T}}.
\tag{2.159}
$$

For computing the 2-D transforms, we apply 1-D transforms to all rows and to all columns successively in Eq. (2.160).

$$
\begin{aligned}
\mathbf{F} &= \Phi \mathbf{f} \Psi^{\mathbf{T}}, \\
&= \Phi \left( \left( \Psi \mathbf{f}^{\mathbf{T}} \right)^{*} \right).
\end{aligned}
\tag{2.160}
$$

Definitions of all the transforms discussed in 1-D are trivially extended to 2-D following the framework as presented above. Most of their properties in 1-D also could be extended to 2-D. However, as DCT is used in compression standards, we review some of its properties in 2-D in the following subsection.

## 2.3.1 2-D Discrete Cosine Transform

In 2-D also there are *eight* different types of DCT. Out of them, we concentrate only on the type-I and type-II *even* 2-D DCTs . For an input sequence $x(m, n), m = 0, 1, 2, \ldots, M; n = 0, 1, 2, \ldots, N$, these are defined as follows:

$$X_I(k, l) = \frac{2}{N}.\alpha^2(k).\alpha^2(l). \sum_{m=0}^{M} \sum_{n=0}^{N} (x(m, n) \cos(\frac{m\pi k}{M}) \cos(\frac{n\pi l}{N})),$$
$$0 \le k \le M, 0 \le l \le N. \tag{2.161}$$

$$X_{II}(k, l) = \frac{2}{N}.\alpha(k).\alpha(l). \sum_{m=0}^{M-1} \sum_{n=0}^{N-1} (x(m, n) \cos(\frac{(2m+1)\pi k}{2M}) \tag{2.162}$$
$$\cos(\frac{(2n+1)\pi l}{2N})), \ 0 \le k \le M - 1, 0 \le l \le N - 1.$$

The type-I 2-D DCT is defined over $(M+1) \times (N+1)$ samples, whereas the type-II 2-D DCT is defined over $M \times N$ samples. These can also be derived from the 2-D GDFT of symmetrically extended sequences, as in the 1-D case. We denote the type-I and the type-II 2-D DCTs of $x(m, n)$ by $C_{1e}\{x(m, n)\}$ and $C_{2e}\{x(m, n)\}$, respectively.

### 2.3.1.1 Matrix Representation

A 2-D input sequence $\{x(m, n), 0 \le m \le M - 1, 0 \le n \le N - 1\}$ is represented by $M \times N$ matrix **x**. Its DCT is expressed in the following form:

$$X = DCT(x) = C_M.\mathbf{x}.C_N^T. \tag{2.163}$$

### 2.3.1.2 Subband Approximation of the Type-II DCT

The approximate DCT computation scheme as discussed earlier, exploiting subband relationship of type-II DCT coefficients, can also be directly extended to the 2-D. Let the *low–low* subband $x_{LL}(m, n)$ of the image $x(m, n)$ be computed as

$$x_{LL}(m, n) = \frac{1}{4}\{x(2m, 2n) + x(2m + 1, 2n)$$
$$+ x(2m, 2n + 1) + x(2m + 1, 2n + 1)\}, \ 0 \le m, n \le \frac{N}{2} - 1. \tag{2.164}$$

Let $\overline{X_{LL}}(k, l), 0 \le k, l \le \frac{N}{2} - 1$ be the 2D DCT of $x_{LL}(m, n)$. Then the *subband approximation* of the DCT of $x(m, n)$ is given by [72]

$$X(k, l) = \begin{cases} 2\cos(\frac{\pi k}{2N}) \cos(\frac{\pi l}{2N}) \overline{X_{LL}}(k, l), & k, l = 0, 1, \ldots, \frac{N}{2} - 1, \\ 0, & \text{otherwise.} \end{cases} \tag{2.165}$$

Similarly, the *low-pass truncated approximation* of the DCT is given by

$$X(k, l) = \begin{cases} 2\overline{X_{LL}}(k, l), & k, l = 0, 1, \ldots, \frac{N}{2} - 1, \\ 0, & \text{otherwise.} \end{cases} \tag{2.166}$$

**2.3.1.3   Composition and Decomposition of the DCT Blocks in 2-D**

Following the same analysis as presented for 1-D [67], we express the block composition and decomposition operations in 2-D. Let $A_{L,N}$ be the block composition matrix, which combines $L$ $N$-point DCT blocks into a block of $LN$-point DCT. Consider $L \times M$ number of $N \times N$-DCT blocks in 2-D. Then its composition into a single $LN \times MN$ block is expressed as

$$X^{(LN \times MN)} = A_{(L,N)} \begin{bmatrix} X_{0,0}^{(N \times N)} & X_{0,1}^{(N \times N)} & \cdots & X_{0,M-1}^{(N \times N)} \\ X_{1,0}^{(N \times N)} & X_{1,1}^{(N \times N)} & \cdots & X_{1,M-1}^{(N \times N)} \\ \vdots & \vdots & \ddots & \vdots \\ X_{L-1,0}^{(N \times N)} & X_{L-1,1}^{(N \times N)} & \cdots & X_{L-1,M-1}^{(N \times N)} \end{bmatrix} A_{(M,N)}^T .$$

(2.167)

Similarly, for decomposing a DCT block $X^{(LN \times MN)}$ to $L \times M$ DCT blocks of size $N \times N$ each, the following expression is used:

$$\begin{bmatrix} X_{0,0}^{(N \times N)} & X_{0,1}^{(N \times N)} & \cdots & X_{0,M-1}^{(N \times N)} \\ X_{1,0}^{(N \times N)} & X_{1,1}^{(N \times N)} & \cdots & X_{1,M-1}^{(N \times N)} \\ \vdots & \vdots & \ddots & \vdots \\ X_{L-1,0}^{(N \times N)} & X_{L-1,1}^{(N \times N)} & \cdots & X_{L-1,M-1}^{(N \times N)} \end{bmatrix} = A_{(L,N)}^{-1} X^{(LN \times MN)} A_{(M,N)}^{-1^T} .$$

(2.168)

**2.3.1.4   Symmetric Convolution and Convolution–Multiplication Properties for 2-D DCT**

Like 1-D, similar convolution–multiplication properties also hold here. In particular, operations involving both type-I and type-II even DCTs are stated below [95].

$$C_{2e}\{x(m,n) \circledast h(m,n)\} \quad = \quad C_{2e}\{x(m,n)\}C_{1e}\{h(m,n)\}, \qquad (2.169)$$

$$C_{1e}\{x(m,n) \circledast h(m,n)\} \quad = \quad C_{2e}\{x(m,n)\}C_{2e}\{h(m,n)\}. \qquad (2.170)$$

Equations (2.169) and (2.170) involve $M \times N$ multiplications for performing the convolution operation in the transform domain.

**2.3.1.5   Fast DCT Algorithms**

As observed earlier, it is not sufficient to propose or design an equivalent algorithm in the compressed domain. We should also consider the merits and demerits of the scheme with respect to alternative spatial domain processing. A faster DCT computation makes spatial domain processing more attractive as the overhead of forward and inverse transforms gets reduced subsequently. In this section let us review a few of these efficient approaches for computing the DCT coefficients. Various algorithms are reported for computing DCT of a sequence efficiently [3, 24, 25, 26, 37, 38, 41, 74, 88, 146, 157]. Out of them, a few demand special architecture for processing the data. Duhamel and Mida [38] provided a theoretical lower bound on the multiplicative complexity of the 1-D $N(= 2^n)$-point DCT as

$$\mu(C_N) = 2^{n+1} - n - 2, \tag{2.171}$$

where the length of the DCT is $N = 2^n$, and $\mu$ denotes the minimum number of multiplications required in the computation. Loeffler et al. [88] proposed a fast 1-D 8-point DCT algorithm with 11 multiplications and 29 additions using graph transformations and equivalence relations. This algorithm achieves the theoretical lower bound on the required number of multiplications (that is, 11 for 1-D DCT).

In 2-D, algorithms using polynomial transforms [37, 146] are shown to be very efficient. These algorithms require a comparable number of additions and the smallest number of multiplications of all known algorithms, leading to an almost 50% reduction in multiplications required by the conventional fast separable DCT computation. However, polynomial algorithms are more complex for implementation. Cho and Lee [26] introduced a direct 2-D $4 \times 4$-point DCT algorithm achieving similar computational complexity, but maintaining simple and regular computational structures typical of the direct row–column and the 2-D vector radix DCT algorithms [158]. This algorithm computes the 2-D DCT from four 1-D 4-point DCTs [25]. For determining the number of multiplications and additions they considered applying the fast 1-D DCT algorithm either by Lee [82] or Hou [59]. Thus, their algorithm requires 96 multiplications and 466 additions for the 2-D $8 \times 8$-point DCT.

The theoretical bound of the multiplicative complexity in 2-D is determined by Feig and Winograd [41]. For a 2-D $2^n \times 2^n$-point DCT, it is given as

$$\mu(C_N \otimes C_N) = 2^n(2^{n+1} - n - 2), \tag{2.172}$$

where the length of the DCT is $N = 2^n$ and $\otimes$ is the Kronecker product and $C_N \otimes C_N$ represents the 2-D DCT matrix. They proposed an algorithm for computing the $8 \times 8$ block DCT, which requires 94 multiplications and 454 additions. Subsequently, Wu and Man [157] pointed out that, using the fast 1-D 8-point DCT algorithm [88] in Cho and Lee's technique for computing 2-D $8 \times 8$ DCT [25], we may achieve the optimal number of multiplications. The resulting number of multiplications is $8 \times 11 = 88$, which is the same as the multiplicative lower bound of Eq (2.172). The number of additions remains the same as before (466).

However, it is possible to compute even with less number of multiplications if we uses other different operations such as shift, 2's complement operations, etc. For example, in [5] and [78], two such algorithms are reported that require only 5 multiplications, 29 additions, and a few two's complement operations (reportedly 16 in [5] and 12 in [78]) for computing an 8-point DCT. Thus, according to [78], only 80 multiplications, 464 additions, and 192 two's complement operations are required for computing the 2-D $8 \times 8$-point coefficients. The complexities of various $8 \times 8$ DCT computation algorithms are summarized in Table 2.9.

Table 2.9: Complexity comparison of various algorithms for $8 \times 8$ DCT

| Algorithm | Year | 1-D DCT | | 2-D DCT | |
|---|---|---|---|---|---|
| | | $\mathcal{M}$ | $\mathcal{A}$ | $\mathcal{M}$ | $\mathcal{A}$ |
| Ahmed et al. [3] | 1974 | 64 | 64 | 1024 | 1024 |
| Chen et al. [24] | 1977 | 16 | 26 | 224 | 416 |
| Kamanagar and Rao [74] | 1982 | | | 128 | 430 |
| Vetterli [146] | 1985 | | | 104 | 474 |
| Arai et al. [5] [††] | 1988 | 5 | 29 | 80 | 464 |
| Loeffler et al. [88] | 1989 | 11 | 29 | 176 | 464 |
| Feig and Winograd [41] | 1992 | | | 94 | 454 |
| Cho et al. [26] | 1993 | | | 96 | 466 |
| Wu and Man [157] | 1998 | | | 88 | 466 |

[††] It also requires additional two's complements operations
Source: From [147] with permission from the author.

## 2.3.2    2-D Discrete Wavelet Transform

As discussed for separable transforms, the 2-D DWT also can be implemented by applying the 1-D DWT along the rows and columns of an image. The computational steps are described in Figure 2.8. In this figure, the first level of decomposition or analysis of an input image is shown. As a result of this processing, four sets of wavelet coefficients are obtained. They are referred to as approximation ($LL$), horizontal ($HL$), vertical ($LH$), and diagonal ($HH$) coefficient subbands, respectively. As explained in Chapter 1, in our notation, the first letter L or H corresponds to the application of a low-pass ($L$) or high-pass ($H$) filter to the rows, and the second letter refers to the same application to the columns. After filtering, half the samples are dropped by downsampling with a factor of 2 as is done for a 1-D transform. For the next level of dyadic decomposition, the $LL$ subband is decomposed into four subbands in the same way, and the process could be iterated till the lowest level of resolution of representing the image. For reconstructing the image (also known as synthesis operation), inverse operations are performed in the reverse order. The reconstruction starts from the lowest level of subbands, and from each lower level, the next higher level image representation (the corresponding $LL$ subband) is formed. In this computation, first the subbands are upsampled by a factor of two with zero insertions in a way similar to the 1-D synthesis

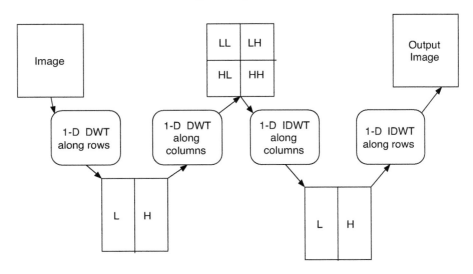

Figure 2.8: 2-D DWT and IDWT.

process. The upsampled subbands are then filtered by corresponding synthesis filters and the results are added to obtain the reconstructed image for the next higher level. In this case also, efficient DWT and IDWT computations can be performed by lifting stages for individual 1-D transformation.

### 2.3.2.1  Computational Complexity

Let us consider the cost of filtering for 1-D DWT with the filters $h(n)$ and $g(n)$. Let the corresponding length of these filters be $|h|$ and $|g|$, respectively. For every output sample, the number of multiplications and additions in convolving an FIR filter of length $l$ are $(l+1)$ and $l$, respectively. Hence, for every pair of wavelet subband coefficients $((a(n), d(n)))$, we require $|h| + |g| + 2$ multiplications and $(|h| + |g|)$ additions. However, if the filters are symmetric, the number of multiplications gets reduced to $(\lceil \frac{|h|}{2} \rceil + \lceil \frac{|g|}{2} \rceil + 2)$. Hence, for implementing the 5/3 filter, we require 7 multiplications and 8 additions for every two samples. In 2-D, these operations are performed twice for every pixel in the reconstructed image. Thus, the number of per-pixel operations in this case is 7 multiplications and 8 additions $(7M + 8A)$. Similarly, 9/7 filtering can be performed with 13 multiplications and 16 additions for every pixel in 2-D. These computations can be performed much faster using lifting schemes. Daubechies discussed in [32] that for even $|h|$ and $|g|$, the number of operations for 1-D lifting implementation of these filters involve $(\frac{|h|}{2} + \frac{|g|}{2} + 2)$ multiplications and that many number of additions. Typically, 9/7 lifting implementation requires 6 multiplications and 4 additions for every pair of output samples. It implies that, in 2-D, per-pixel computation cost becomes $6M + 8A$. However, in 5/3 lifting, multiplication can be performed by right-shift operations. Hence, it requires only 4 additions and 2 shift operations ($\approx 4A$) for every output pixel in 2-D.

## 2.4   Summary

Image transforms provide alternative descriptions of images and are useful for image and video compression. There are various techniques for transforming or factorizing an image as a linear combination of orthogonal basis functions. For a finite discrete image, the number of these discrete basis functions (or basis matrices) is also finite, and transformations are invertible. Some of the properties of these transforms, in particular, properties of discrete Fourier transforms, generalized discrete Fourier transforms, trigonometric transforms, discrete cosine transforms, and discrete wavelet transforms were reviewed in this chapter.

# Chapter 3

## Image Filtering

Filtering of an image or any signal, in general, is the process of emphasizing or attenuating some of its frequency components. Subsequently, the resulting image or signal retains features associated with those frequency components. Image filtering is often required in various applications such as reducing noise, extracting edges and boundaries of objects, interpolating functional values, etc. The term *filtering* itself is self-explanatory if we consider the representation of a function $f(x)$ in the frequency domain by its *Fourier transform*, say $\widehat{f}(j\omega)$. Given a filtering function $\widehat{h}(j\omega)$, where the frequency component at a frequency $\omega_0$ is modified by a factor of $\widehat{h}(j\omega_0)$, the filtered output in the frequency domain becomes $\widehat{g}(j\omega) = \widehat{h}(j\omega)\widehat{f}(j\omega)$. Hence, the corresponding processed function in its original domain (say, temporal or spatial domain) becomes the *inverse Fourier transform* of $\widehat{g}(j\omega)$, which is represented in the text as $g(x)$. The filtering function in the frequency space ( $\widehat{h}(j\omega)$ in this example) is more commonly referred to as the *frequency response* or *transfer function* of the filter. The equivalent operation of filtering in the spatial or temporal domain (original domain of the function) is the convolution of the function with $h(x)$, which is the inverse Fourier transform of $\widehat{h}(j\omega)$. This has been already discussed in the previous chapter, where the *convolution–multiplication properties* of *Fourier transform* as well as of *discrete Fourier transform* are discussed (see Sections 2.1.3.1 and 2.2.1.3). However, in this chapter, we briefly review this concept with respect to the analysis of a *linear system*, in particular a *discrete linear system*. This would be followed by adaptation of this computation in the block DCT domain, which is different from the frequency domain of the Fourier transform. Convolution–multiplication properties (CMP) with certain restrictions are also applicable to *trigonometric transforms*, as discussed before (see Chapter 2, Section 2.2.3). We also discuss their suitable adaptation for filtering images in this chapter. Here, initially, we limit our discussion to 1-D only. Later the concepts are extended to 2-D.

## 3.1   Linear Shift Invariant (LSI) Systems

**Definition 3.1 Linear system:** *A system is said to be* linear *if a* linear combination *of inputs provide the* same linear combination *of their outputs. For example, if responses for inputs $x_1$ and $x_2$ are $y_1$ and $y_2$, respectively, the response of $ax_1 + bx_2$, where $a$ and $b$ are arbitrary scalar constants, is $ay_1 + by_2$.*   □

In our discussion we consider systems processing either time- or space-varying signals into similar forms and thus characterizing them in terms of their input–output responses.

**Definition 3.2 Shift invariant system:** *A system is* shift invariant *if the response of a translated input in its domain maintains the same translation in its original response. For example, if the response of $f(x)$ is $g(x)$, for a shift invariant system, the response of $f(x - x_0)$ is $g(x - x_0)$ for all values of $x_0$.*

□

   A system is called *linear shift invariant* (LSI), when it is both linear and shift invariant. For an LSI system, we characterize its behavior from its *impulse response*, which is the response obtained from the system given a unit infinite impulse (modeled as the *Dirac delta* function $\delta(x)$) as its input. This is stated in the following theorem.

**Theorem 3.1** *Given the* impulse response *of an LSI system as $h(x)$, the* output response *$g(x)$ of an input $f(x)$ is obtained by the following expression.*

$$g(x) = \int_{-\infty}^{\infty} f(\chi)h(x - \chi)d\chi. \tag{3.1}$$

*The above operation is also known as the* convolution *operation and denoted by the operator '★' such that $g(x) = h \star f(x)$.* □

The convolution operation is *commutative* but *not associative*. Moreover, for a *bounded* function, we get a *bounded* response if and only if its impulse response satisfies the following condition.

$$\int_{-\infty}^{\infty} |h(x)| dx < \infty. \tag{3.2}$$

We also discussed in the previous chapter that the equivalent operation of convolution in the *frequency domain* is *multiplication*, as restated here in the following form.

$$\mathbb{F}(h \star f(x)) = \hat{h}(j\omega)\hat{f}(j\omega). \tag{3.3}$$

The above relationship provides the prime motivation for designing *filters* in the *frequency space* by specifying its frequency response $\hat{h}(j\omega)$ in terms of its *magnitude* ($|\hat{h}(j\omega)|$) and *phase* ($\angle\hat{h}(j\omega)$) at varying frequency. In this case, the computation of the output is usually performed in the frequency space followed by the *inverse Fourier transform* to represent the response in its original domain. However, for *discrete LSI systems*, filters are more often designed in the original space itself by specifying its *impulse response*, and computations are usually performed in the spatial or temporal domain.

## 3.2 Discrete LSI Systems

A *discrete LSI system* processes *discrete functions* in the form of a sequence $f(n), n \in \mathbb{Z}$ as defined earlier. The system has the same linear and shift-invariant properties and can be characterized by the *impulse response*, which is also a sequence (say, $h(n)$) at discrete points in its domain sampled with the same periodicity. Hence, the convolution operation of Eq. ( 3.1) is expressed in the discrete domain in the following form:

$$h \star f(n) = \sum_{m=-\infty}^{\infty} f(m)h(n-m) = \sum_{m=-\infty}^{\infty} h(m)f(n-m). \tag{3.4}$$

The foregoing equation provides *bounded* sequence for a *bounded* input sequence if $h(n)$ satisfies the following condition:

$$\sum_{n=-\infty}^{\infty} |h(n)| < \infty. \qquad (3.5)$$

An impulse response $h(n)$ is *finite* length if $\exists\ N_1, N_2 \in \mathbb{Z}$ such that $h(k) = 0, \forall k \leq N_1$ and $\forall k \geq N_2$. Otherwise, the impulse response is called *infinite length impulse response*. For a *finite length impulse response* $h(n)$, the system is *stable* if $|h(n)| < \infty, \forall n \in \mathbb{Z}$. In our context, we refer to a discrete LSI system as a *filter* and its impulse response as the *filter response*. A filter with *finite length impulse response* is called a *finite impulse response (FIR) filter*, whereas a filter with *infinite length impulse response* is known as an *infinite impulse response (IIR) filter*. As in most of the applications of image and video processing *FIR filters* are used, we restrict our discussion to this type of filter.

---

## 3.3   Filtering a Finite Length Sequence

Let us consider the filtering of a finite length sequence $x(n), 0 \leq n \leq N - 1$ of length $N$. Given a finite length impulse response $h(n)$, we investigate the relationship between the output and input of an LSI system. As the finite length sequence is defined only for the interval between 0 and $N - 1$, let us consider how to exploit this fact by defining functional values at other undefined sample points for modeling certain real-world input conditions or for satisfying certain constraints in computation. The observation window (the interval for the output response) is also the other flexibility in designing this computational model. In the following subsections we discuss the different types of extended definition of the finite length sequence and the corresponding computation for obtaining the output response from the system.

### 3.3.1   Extension by Zero Padding

This is the usual notion of a finite length sequence by considering $x(n) = 0$ for $n \geq N$ and $n < 0$. In this case, given an impulse response $h(n)$, the *linear convolution* as given in Eq. (3.4) is applicable. The corresponding CMP holds in the z-transform [100] domain in this case, which is stated in the following theorem.

**Theorem 3.2** *Let the z-transform of $x(n)$ and $h(n)$ be $X(Z)$ and $H(Z)$, respectively. In that case, the z-transform of $y(n) = x \bigstar h(n)$ is obtained as*

$$Y(Z) = H(Z)X(Z), \qquad (3.6)$$

*where $Y(Z)$ is the z-transform of $y(n)$.* □

### 3.3.1.1 Linear Convolution Matrix

Let us consider a type of FIR filter with an impulse response $h(n), 0 \leq n \leq L-1$ of length $L$. It is observed that the output response of this filter has a length of $N+L-1$ given the input of length $N$. Accordingly, by considering the observation window between 0 and $N+L-1$, the input–output relationship is expressed by the following matrix representation. In this case, as before, a finite length sequence is represented by a column vector.

$$
\begin{bmatrix} y(0) \\ y(1) \\ \cdot \\ \cdot \\ y(L-1) \\ \cdot \\ y(N-1) \\ \cdot \\ y(N+L-1) \end{bmatrix} = H_l^{(N+L-1)\times N} \begin{bmatrix} x(0) \\ x(1) \\ \cdot \\ \cdot \\ x(L-1) \\ \cdot \\ x(N-1) \end{bmatrix}. \tag{3.7}
$$

where,

$$
H_l^{(N+L-1)\times N} = \begin{bmatrix} h(0) & 0 & \cdots & 0 & \cdots & 0 \\ h(1) & h(0) & \cdots & 0 & \cdots & 0 \\ \cdot & \cdot & \cdots & \cdot & \cdots & \cdot \\ \cdot & \cdot & \cdots & \cdot & \cdots & \cdot \\ h(L-1) & h(L-2) & \cdots & h(0) & \cdots & 0 \\ \cdot & \cdot & \cdots & \cdot & \cdots & \cdot \\ 0 & 0 & \cdots & \cdot & \cdots & h(0) \\ \cdot & \cdot & \cdots & \cdot & \cdots & \cdot \\ 0 & 0 & \cdots & 0 & \cdots & h(L-1) \end{bmatrix}. \tag{3.8}
$$

In the above example, we have considered $L$ to be less than $N$. Given a finite length filter response $h(n)$, we obtain the *linear convolution matrix* (denoted as $H_l^{(N+L-1)\times N}$) in Eq. (3.7). The dimension of this matrix is $(N+L-1)\times N$. Due to the linear and distributive property of DCTs, we express Eq. (3.7) in the transform domain as follows:

$$
DCT(\mathbf{y}) = DCT(H_l^{(N+L-1)\times N})DCT(\mathbf{x}), \tag{3.9}
$$

where $DCT(.)$ denotes the appropriate DCT (in 1-D or 2-D with proper dimensionality of a vector or matrix), and $\mathbf{y}$ and $\mathbf{x}$ denote the output and input column vectors, respectively.

### 3.3.2    Periodic Extension

The periodic extension of $x(n)$ with a period $N$ implies that $x(N+n) = x(n)$. In this case, a filter with impulse response with the same periodicity $N$ generates an output with the periodicity $N$. With this property, we also consider the periodic extension of an FIR filter with a response $h(n), 0 \leq n \leq N-1$ of length $N$. By keeping the observation window between 0 and $N-1$, the output of this system is obtained by the *circular convolution* as discussed in the previous chapter. The operator for this operation is denoted in this text as $\circledast$, and the operation is shown below.

$$y(n) = x(n) \circledast h(n) = \sum_{m=0}^{n} x(m)h(n-m) + \sum_{m=n+1}^{N-1} x(m)h(n-m+N). \quad (3.10)$$

It has been discussed in the previous chapter that the *convolution–multiplication* property for the *circular convolution* holds in the discrete Fourier transform space. This is restated here in the following theorem.

**Theorem 3.3** *Let DFT of $x(n)$ and $h(n)$ be $\widehat{x}(k)$ and $\widehat{h}(k)$, respectively. In that case, the DFT of $y(n) = x(n) \circledast h(n)$ is obtained as*

$$\widehat{y}(k) = \widehat{h}(k)\widehat{x}(k), \quad (3.11)$$

*where $\widehat{y}(k)$ is the DFT of $y(n)$.*                    □

#### 3.3.2.1    Circular Convolution Matrix

As in the case of linear convolution we express the computation of circular convolution with a convolution matrix (say, $H_c^{N \times N}$) in the following form.

$$\begin{bmatrix} y(0) \\ y(1) \\ \cdot \\ \cdot \\ y(N-1) \end{bmatrix} = H_c^{N \times N} \begin{bmatrix} x(0) \\ x(1) \\ \cdot \\ \cdot \\ x(N-1) \end{bmatrix}, \quad (3.12)$$

where,

$$H_c^{N \times N} = \begin{bmatrix} h(0) & h(N-1) & \cdots & h(1) \\ h(1) & h(0) & \cdots & h(2) \\ \cdot & \cdot & \cdots & \cdot \\ \cdot & \cdot & \cdots & \cdot \\ h(N-1) & h(N-2) & \cdots & h(0) \end{bmatrix}. \quad (3.13)$$

From Eq. ( 3.3) we find that the $N$-point DFT matrix diagonalizes the circular convolution matrix. Hence, the basis vectors of the corresponding DFT become the *eigenvectors* [55], and the *DFT coefficients* of $h(n)$ are the *eigen-values* [55] of this convolution matrix. Due to the linear and distributive property of DCTs, a relationship similar to Eq. (3.9) also holds here. But it is less efficient to compute with DCT coefficients in this form than that with DFT coefficients.

### 3.3.2.2 Linear Convolution Performed through Circular Convolution

As linear convolution is the requirement for processing an input sequence with an *FIR* filter, it is of interest to adapt the circular convolution operation to perform similar task. In this case, given an impulse response $h(n)$ of length $L$ and an input sequence $x(n)$ of length $N$, the sequence is padded with adequate number of trailing zeroes to both these sequences so that their length becomes equal. The circular convolution of these zero-padded sequences provides a result similar to that is obtained by the linear convolution of $x(n)$ and $h(n)$. In the previous subsection we have observed that the length of the output for linear convolution under this circumstance becomes $N + L - 1$. Hence, in this case, the number of trailing zeroes to be added for $h(n)$ and $x(n)$ should be at least $N - 1$ and $L - 1$, respectively. With this extension, the circular convolution between them (both of length $N+L-1$) produces the same output sequence. Hence, we require $N + L - 1$-point DFT to perform the equivalent operation in the frequency domain.

### 3.3.3 Antiperiodic Extension

In an *antiperiodic extension*, $x(n)$ is defined outside the interval between 0 and $N-1$ as $x(N+n) = -x(n)$, where $N$ is the period of *antiperiodicity*. The extended signal has the *strict periodicity* of $2N$. In this case, if a finite impulse response $h(n), 0 \leq n \leq N-1$ is extended antiperiodically in the same way, its convolution with the extended $x(n)$ produces an output with the same $(N)$ antiperiodicity. By keeping the observation window restricted between 0 and $N - 1$, the convolution expression is expressed in the following form.

$$y(n) = x(n) \circledS h(n) = \sum_{m=0}^{n} x(m)h(n-m) - \sum_{m=n+1}^{N-1} x(m)h(n-m+N). \quad (3.14)$$

This type of convolution is called *skew circular convolution* and is denoted by the operator $\circledS$. In this case, the CMP is true for *odd frequency discrete Fourier transform (OFDFT)* (see Section 2.2.2 of Chapter 2), which is restated in the following form.

**Theorem 3.4** *Let $\widehat{x}_{\frac{1}{2},0}(k)$ and $\widehat{h}_{\frac{1}{2},0}(k)$ be the OFDFTs of $x(n)$ and $h(n)$, respectively. The OFDFT of their skew circular convolved output $y(n) = x(n)\circledS h(n)$ is obtained by the following expression.*

$$\widehat{y}_{\frac{1}{2},0}(k) = \widehat{h}_{\frac{1}{2},0}(k)\widehat{x}_{\frac{1}{2},0}(k), \quad (3.15)$$

*where $\widehat{y}_{\frac{1}{2},0}(k)$ is the OFDFT of $y(n)$.* $\square$

### 3.3.3.1    Skew Circular Convolution Matrix

As before, we also express the convolution operation with the help of matrix operations and denote the corresponding convolution matrix as $H_s^{N \times N}$. The convolution matrix takes the following form:

$$H_s^{N \times N} = \begin{bmatrix} h(0) & -h(N-1) & \cdots & -h(1) \\ h(1) & h(0) & \cdots & -h(2) \\ \cdot & \cdot & \cdots & \cdot \\ \cdot & \cdot & \cdots & \cdot \\ h(N-1) & h(N-2) & \cdots & h(0) \end{bmatrix}. \tag{3.16}$$

The *skew circular convolution* matrix is obtained from the circular convolution matrix by negating its elements above the main diagonal.

### 3.3.3.2    Circular Convolution as a Series of Skew Circular Convolution

In [110] it is shown that circular convolution is performed by a series of skew circular convolution by partitioning the sequence at each stage and, subsequently, the output is obtained as a concatenation of results from individual partitions. Let us consider the partitioning of the input sequence $x(n)$ of length $N$ into two $\frac{N}{2}$ halves $x_1(n), 0 \leq n \leq \frac{N}{2} - 1$, and $x_2(n), \frac{N}{2} \leq n \leq N - 1$. Now, Eq. (3.12) is rewritten in the following form.

$$\begin{bmatrix} \mathbf{y_1} \\ \mathbf{y_2} \end{bmatrix} = \begin{bmatrix} A & B \\ B & A \end{bmatrix} \begin{bmatrix} \mathbf{x_1} \\ \mathbf{x_2} \end{bmatrix}. \tag{3.17}$$

In the above equation, $A$ and $B$ are submatrices of $H_c^{N \times N}$ of dimension $\frac{N}{2} \times \frac{N}{2}$, and $\mathbf{y_1}$ and $\mathbf{y_2}$ are subcolumn vectors corresponding to two halves of the output response. Eq. (3.17) is also written in the following form:

$$\begin{array}{rcl} \mathbf{y_1} & = & \frac{1}{2}(A+B)(\mathbf{x_1}+\mathbf{x_2}) + \frac{1}{2}(A-B)(\mathbf{x_1}-\mathbf{x_2}), \\ \mathbf{y_2} & = & \frac{1}{2}(A+B)(\mathbf{x_1}+\mathbf{x_2}) - \frac{1}{2}(A-B)(\mathbf{x_1}-\mathbf{x_2}). \end{array} \tag{3.18}$$

It can be shown that $\frac{1}{2}(A+B)$ and $\frac{1}{2}(A-B)$ are circular convolution and skew-circular convolution matrices, respectively. This process is iterated by partitioning $\mathbf{y_1}$ in the same way till we get a trivial circular convolution matrix (of dimension $2 \times 2$). In [110], an efficient computation of skew circular convolution of a real sequence is described by using the *generalized DFT*.

### 3.3.4    Symmetric Extension

In Section 2.2.3.1 of chapter 2, we have discussed the different types of symmetric extension of a finite sequence and their implications in definitions of various discrete trigonometric transforms such as *type-I to type-IV even and odd DCTs* as well as DSTs. We have also listed several CMPs as applicable

**Table 3.1**: Convolution–multiplication properties for symmetrically extended sequences involving *type-II even DCT* and *type-II even DST*

| Symmetric extensions | | Ranges of | | | Convolution–multiplication |
|---|---|---|---|---|---|
| $h(n)$ | $x(n)$ | $h(n)$ | $x(n)$ | $y(n) =$ $\tilde{x}(n) \circledast \tilde{h}(n)$ | property |
| WSWS | HSHS | $0 \rightarrow N$ | $0 \rightarrow N-1$ | $0 \rightarrow N-1$ | $C_{2e}(y(n)) = \sqrt{2N}C_{1e}(h(n))C_{2e}(x(n))$ |
| WSWS | HAHA | $0 \rightarrow N$ | $0 \rightarrow N-1$ | $0 \rightarrow N-1$ | $S_{2e}(y(n)) = \sqrt{2N}C_{1e}(h(n))S_{2e}(x(n))$ |
| WAWA | HSHS | $1 \rightarrow N$ | $0 \rightarrow N-1$ | $0 \rightarrow N-1$ | $S_{2e}(y(n)) = \sqrt{2N}S_{1e}(h(n))C_{2e}(x(n))$ |
| WAWA | HAHA | $1 \rightarrow N$ | $0 \rightarrow N-1$ | $0 \rightarrow N-1$ | $-C_{2e}(y(n)) = \sqrt{2N}S_{1e}(h(n))S_{2e}(x(n))$ |

to different pairs of symmetrically extended sequences. In this chapter we observe the structure and properties of convolution matrices for various pairs of symmetrically extended sequences. Later on we explore their applicability in designing algorithms for filtering in the block DCT domain. In our discussion we restrict our attention to those convolution–multiplication properties that involve either a *type-II even DCT* or a *type-II even DST*. The corresponding types of extensions and their CMPs are shown in Table 3.1. In the table, the output sequence $y(n)$ is computed as the circular convolution of the extended sequences obtained from $x(n)$ (denoted by $\tilde{x}(n)$) and $h(n)$ (denoted by $\tilde{h}(n)$), respectively. The type of corresponding extensions are also shown in respective entries of the table. We use here same notations for representing different types of *even* DCTs and DSTs as used in the previous chapter (Section 2.2.3.3 of Chapter 2).

### 3.3.4.1 Symmetric Convolution Matrices

Let us study the structure of the convolution matrices for the cases mentioned in Table 3.1. Let us represent the corresponding convolution matrix with the notation $H_{a,b}$ such that $a \in \{WA, WS\}$ and $b \in \{HA, HS\}$ for performing convolution between $h(n)$ and $x(n)$ by extending them according to the symmetry (or antisymmetry) $a$ and $b$ at their both ends, respectively. The dimension of $H_{a,b}$ is $N \times N$, such that

$$\mathbf{y} = H_{a,b}\mathbf{x}, \tag{3.19}$$

where $\mathbf{y}$ and $\mathbf{x}$ are column vectors of length $N$ for the output and input, respectively.

In [79] it has been shown that $H_{a,b}$s can be expressed as linear combination of the following matrices. Let us denote these matrices as $H_1$, $H_2$, $H_3$, and

$H_4$. They are given below.

$$H_1 = \begin{bmatrix} \frac{h(0)}{2} & 0 & \cdots & 0 & 0 \\ h(1) & \frac{h(0)}{2} & \cdots & 0 & 0 \\ \cdot & \cdot & \cdots & \cdot & \cdot \\ \cdot & \cdot & \cdots & \cdot & \cdot \\ h(N-2) & h(N-3) & \cdots & \frac{h(0)}{2} & 0 \\ h(N-1) & h(N-2) & \cdots & h(1) & \frac{h(0)}{2} \end{bmatrix}, \qquad (3.20)$$

$$H_2 = \begin{bmatrix} h(1) & h(2) & \cdots & h(N-1) & \frac{h(N)}{2} \\ h(2) & h(3) & \cdots & \frac{h(N)}{2} & 0 \\ \cdot & \cdot & \cdots & \cdot & \cdot \\ \cdot & \cdot & \cdots & \cdot & \cdot \\ h(N-1) & \frac{h(N)}{2} & \cdots & 0 & 0 \\ \frac{h(N)}{2} & 0 & \cdots & 0 & 0 \end{bmatrix}, \qquad (3.21)$$

$$H_3 = \begin{bmatrix} \frac{h(0)}{2} & h(1) & \cdots & h(N-2) & h(N-1) \\ 0 & \frac{h(0)}{2} & \cdots & h(N-3) & h(N-2) \\ \cdot & \cdot & \cdots & \cdot & \cdot \\ \cdot & \cdot & \cdots & \cdot & \cdot \\ 0 & 0 & \cdots & \frac{h(0)}{2} & h(1) \\ 0 & 0 & \cdots & 0 & \frac{h(0)}{2} \end{bmatrix}, \qquad (3.22)$$

$$H_4 = \begin{bmatrix} 0 & 0 & \cdots & 0 & \frac{h(N)}{2} \\ 0 & 0 & \cdots & \frac{h(N)}{2} & h(N-1) \\ \cdot & \cdot & \cdots & \cdot & \cdot \\ \cdot & \cdot & \cdots & \cdot & \cdot \\ 0 & \frac{h(N)}{2} & \cdots & h(3) & h(2) \\ \frac{h(N)}{2} & h(N-1) & \cdots & h(2) & h(1) \end{bmatrix}. \qquad (3.23)$$

Finally, $H_{a,b}$s are expressed by foregoing matrices as follows:

$$\begin{aligned} H_{WS,HS} &= (H_1 + H_3) + (H_2 + H_4), \\ H_{WA,HS} &= (H_1 - H_3) + (H_2 - H_4), \\ H_{WS,HA} &= (H_1 + H_3) - (H_2 + H_4), \\ H_{WA,HA} &= (H_1 - H_3) - (H_2 - H_4). \end{aligned} \qquad (3.24)$$

We find that $H_3 = H_1^T$. Both $H_1 + H_3$ and $H_1 - H_3$ are *Toeplitz matrices*[1]. However, the former is a *symmetric* matrix and the latter is an *antisymmetric* one. Similarly, $H_2 + H_4$ and $H_2 - H_4$ are *symmetric and antisymmetric Hankel matrices,*[2] respectively. From the convolution–multiplication properties,

---

[1] Elements along diagonals of a Toeplitz matrix are of the same values.
[2] Elements along off-diagonals of a Hankel matrix are of the same values.

diagonalizations of these matrices are expressed by the following equation.

$$
\begin{array}{rcl}
H_{WS,HS} & = & C^T \mathbb{D}(\sqrt{2N}\{C_{1e}\mathbf{h}\}_0^{N-1})C, \\
H_{WA,HS} & = & S^T \mathbb{D}_1(\sqrt{2N}\{S_{1e}\mathbf{h}\}_1^{N-1})C, \\
H_{WS,HA} & = & S^T \mathbb{D}(\sqrt{2N}\{C_{1e}\mathbf{h}\}_1^{N})S, \\
H_{WA,HA} & = & -C^T \mathbb{D}_{-1}(\sqrt{2N}\{S_{1e}\mathbf{h}\}_1^{N-1})S.
\end{array}
\tag{3.25}
$$

where $C$ and $S$ are the type-II even DCT and DST matrices, respectively. Similarly, $C_{1e}$ and $S_{1e}$ are the respective type-I even DCT and DST matrices. $\mathbb{D}(\mathbf{x})$ denotes the *diagonalization* of a column vector $\mathbf{x}$ of length $N$ into a diagonal matrix of $N \times N$ whose diagonal element at $i$th row is $x(i)$. Similarly, $\mathbb{D}_m(.)$ denotes a square matrix generated from its input column vector such that its $m$th *off-diagonal elements*[3] are formed from it. Finally, $\{\mathbf{x}\}_p^q$ denotes the column vector formed from $\mathbf{x}$ from its $p$th element to the $q$th one.

### 3.3.4.2 Linear Convolution through Symmetric Convolution

*Symmetric convolution* can also be used to perform the linear convolution of a sequence $x(n)$ of length $N$ with an impulse response $h(n)$ of length $L$. However, $h(n)$ should have points of symmetry or antisymmetry so that one half of it (or part of it) generates the complete sequence through symmetric extensions. Types of symmetric extensions at its two ends (of the selected part) as well as the chosen symmetric extensions of the input signal determine the type of convolution to be performed eventually. Under the context of CMPs involving only type-II even DCT and DST (see Table 3.1), impulse responses should have a form such that symmetric extension of type *Whole-sample Symmetric, and Whole-sample Symmetric* (WSWS), or *Whole-sample Antisymmetric, and Whole-sample Antisymmetric* (WAWA) generates it fully. Such an example is shown in Figure 3.1(a) and (b). In this case, the *WSWS* extension applied to the right half of Figure 3.1(a) (i.e., Figure 3.1(b) itself) generates itself. Hence, for performing equivalent linear convolution, we need to pad zeroes to the input sequence (say, as shown in Figure 3.1(c)) not only to its end but also to its beginning. Given the length of the impulse response as $L$, we require $\frac{L+1}{2}$ number of zeroes to be added at both its ends. Similarly, for the *WAWA* type of extension to the right half of an impulse response (and thus generating the full response through this extension), it requires $\frac{L-1}{2}$ number of zeroes to be added at both ends of the input data for obtaining the same result of a linear convolution. We present here in Table 3.2 from [95] the full list of number of zeroes required to be added at both ends for every type of symmetric extension. Depending on the type of convolution, which is again determined by the types of symmetric extensions to the filter response as well as to the input sequence, we should apply the appropriate *convolution–*

---

[3]The $m$th off-diagonal is formed by the $(i, j)$th elements of the matrix such that $j-i = m$. An element $x(k)$ of the vector $\mathbf{x}$ is inserted into the $k$th appearance of a matrix-location satisfying the above property while scanning its rows and columns from left to right and top to bottom.

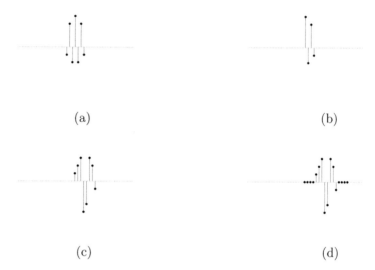

(a)                    (b)

(c)                    (d)

**Figure 3.1**: (a) Impulse response, (b) right-half response, (c) input sequence, and (d) zero-padded input sequence.

*multiplication property* (see Table 3.1). In this computation, we have to take care of the boundary conditions of the input data and adapt it accordingly before applying the transformations. The preceding and trailing zeroes of input data provide the equivalent condition of a finite length sequence for a linear convolution. However, this may not be ideal for a block-based computation. We elaborate on this in the next section.

**Table 3.2**: Number of zeroes required to be added at both ends for every type of symmetric extensions

| Left end symmetry | Right end symmetry | | | |
|---|---|---|---|---|
| | WS | WA | HS | HA |
| WS | $\frac{L+1}{2}$ | $\frac{L-1}{2}$ | $\frac{L-1}{2}$ | $\frac{L-1}{2}$ |
| WA | $\frac{L+1}{2}$ | $\frac{L-1}{2}$ | $\frac{L-1}{2}$ | $\frac{L-1}{2}$ |
| HS | $\frac{L}{2}$ | $\frac{L}{2}-1$ | $\frac{L}{2}$ | $\frac{L}{2}$ |
| HA | $\frac{L}{2}$ | $\frac{L}{2}-1$ | $\frac{L}{2}$ | $\frac{L}{2}$ |

## 3.4 Block Filtering

Processing input data in blocks is quite common in several applications of image and signal processing. In particular, block processing is quite common in compression of audio, images, and videos. In this type of representation, the advantages lie in keeping the buffer requirement low, introducing inherent parallelism, and opening up opportunity for designing pipelined architecture for the computation. Hence, in this section, we review the approaches for filtering sequences partitioned into blocks of a fixed size (say $M$).

Let us consider filtering of an input sequence $x(n)$ with a finite impulse response $h(n)$. Let the input sequence $x(n), 0 \leq n \leq N-1$ be partitioned into $\frac{N}{M}$ subsequences of length $M$ (assume $N$ is a multiple of $M$). Let the $i$th partition be denoted as $x_i(n), 0 \leq i \leq \frac{N}{M} - 1, iM \leq n \leq (i+1)M - 1$. In a *block filtering*, the objective is to design the *equivalent filtering technique* of the sequence $x(n)$ by processing a block or a set of neighboring blocks. There are two basic approaches for *block filtering*, namely, the *overlapping and add* method and the *overlapping and save* method. In the overlapping and add method, *each block* (say, $x_i$) is independently filtered (say, $y_i(n) = x_i(n) \bigstar h(n)$), assuming that their boundaries in both the directions are padded with zeroes. Finally, the sum total of all such responses provides the same result as obtained by filtering the whole sequence at a time (i.e $y(n) = x(n) \bigstar h(n) = \sum_{i=0}^{\frac{N}{M}-1} y_i(n))$.

For the $i$th block filtering with $h(n)$ of length $L$ generates a sequence $y_i(n)$ of length $M + L - 1$ in between $iM$ and $(i + 1)M + L - 1$.

In the overlapping and save method, for each block (say, the $i$th block), its adjacent neighboring blocks in both the directions are concatenated together in larger block, and this acts as the input to the convolution operation. If the length of adjacent data is sufficient enough to provide the necessary intervals of stimulation affecting its response around every sample point of the central block (that is, the $i$th block), it is sufficient to *save* the response associated with the central block in the resulting output stream. The number of adjacent blocks required for concatenation depends on the length of the impulse response. For a *causal* system, it is sufficient to concatenate only the block to its left. For a length $L$, we need to concatenate at least $\lceil \frac{L}{M} \rceil$ number of blocks from its left. However, for a noncausal response, we need to concatenate from both sides. Suppose $h(n)$ is defined in an interval between $-(L - 1)$ and $(L - 1)$. In that case, the number of blocks to be concatenated from both end is also $\lceil \frac{L}{M} \rceil$.

While designing efficient algorithms, we may restrict computations pertaining to the central block only following both the approaches. Eventually, this makes both the computations equivalent. We discuss next how these approaches are adapted in the transform domain, namely, in the block DCT domain.

### 3.4.1 Overlapping and Save Methods in the Transform Domain

A useful technique under this category was reported by Kresch and Merhav [79]. They have performed filtering of blocks using convolution–multiplication properties for both DCTs and DSTs, and their method is applicable to any arbitrary type of FIR, be it a *symmetric noncausal* or an *asymmetric causal sequence*.

Consider the extended neighborhood of a block $x_i$ with $(i-1)$th and $(i+1)$th block such that the $i$th output block $y_i$ is obtained from the convolution with an FIR $h(n), -N \leq n \leq N$. By matrix notation, we can write it in the following form.

$$y_i = \begin{bmatrix} H_2^+ & (H_1^+ + H_1^-) & H_2^- \end{bmatrix} \begin{bmatrix} x_{i-1} \\ x_i \\ x_{i+1} \end{bmatrix}. \tag{3.26}$$

where $H_1^+ (= H_1^p)$ and $H_2^+ (= H_2^p \Phi_N)$ are formed from the nonnegative half of the impulse response (that is, $h^p(n) = \{h(0), h(1), \cdots, 2h(N)\}$) as shown in Eqs. (3.20) and (3.21), respectively. Similarly, $H_1^- (= H_3^m)$ and $H_2^- (= H_4^m \Phi_N)$ are defined from the nonpositive half of the impulse response (that is, $h^m(n) = \{h(0), h(-1), \cdots, 2h(-N)\}$) following Eqs. (3.22) and (3.23), respectively. $\Phi_N$ is the $N \times N$ flipping matrix and given by the following form.

$$\Phi_N = \begin{bmatrix} 0 & \cdots & 0 & 1 \\ 0 & \cdots & 1 & 0 \\ 0 & \cdots & 0 & 0 \\ 1 & \cdots & 0 & 0 \end{bmatrix}. \tag{3.27}$$

We find that $C_N \Phi_N = \Psi_N C_N$ and $S_N \Phi_N = \Psi_N S_N$, where $\Psi_N = \mathbb{D}(\{(-1)^m\}_{m=0}^{N-1})$.

From Eqs. (3.24) and (3.25), the above relationship is expressed in the transform domain in the following form. Note that we have used both the DCT and DST of input blocks in our expression (see Eq. (3.28)). The DCT of a block $x_i$ is denoted as $X_i^c$, whereas its DST is denoted by $X_i^s$.

$$\begin{aligned} Y_i^c =\ & \tfrac{1}{4}(\mathbb{D}(\sqrt{2N}\{C_{1e}\mathbf{h^p}\}_0^{N-1})(X_i^c + \Psi X_{i-1}^c) \\ & + \mathbb{D}(\sqrt{2N}\{C_{1e}\mathbf{h^p}\}_0^{N-1})(X_i^c + \Psi X_{i-1}^c) \\ & - \mathbb{D}_{-1}(\sqrt{2N}S_{1e}\{\mathbf{h^m}\}_1^{N-1})(X_i^c - \Psi X_{i+1}^c) \\ & + \mathbb{D}_{-1}(\sqrt{2N}S_{1e}\{\mathbf{h^m}\}_1^{N-1})(X_i^c - \Psi X_{i+1}^c) \\ & + C_N S_N^T(\mathbb{D}(\sqrt{2N}\{C_{1e}\mathbf{h^p}\}_1^{N})(X_i^s - \Psi X_{i-1}^s) \\ & + \mathbb{D}_1(\sqrt{2N}S_{1e}\{\mathbf{h^p}\}_1^{N-1})(X_i^c + \Psi X_{i-1}^c) \\ & + \mathbb{D}(\sqrt{2N}\{C_{1e}\mathbf{h^m}\}_1^{N})(X_i^s - \Psi X_{i-1}^s) \\ & + \mathbb{D}_1(\sqrt{2N}S_{1e}\{\mathbf{h^m}\}_1^{N-1})(X_i^c + \Psi X_{i+1}^c))). \end{aligned} \tag{3.28}$$

The above expression could be further simplified for symmetric, antisymmetric, and causal responses. As the input is assumed to be available in the block DCT space, efficient *cosine-to-sine transformation* (CST) and *sine-to-cosine transformation* (SCT) are presented in [79].

In a different approach, Yim [165] used the linear and distributive properties of DCTs for developing efficient algorithms for filters with symmetric response. In this case, the submatrices of the convolution matrix in Eq. (3.26) hold the following relation.

$$H_2^+ = (H_2^-)^T. \tag{3.29}$$

Considering $H_0 = H_1^+ + H_1^-$, Eq. (3.26) is rewritten in the block DCT domain as follows.

$$Y_i^c = DCT(H_2^+)X_{i-1}^c + DCT(H_0)X_i^c + DCT(H_2^-)X_{i+1}^c. \tag{3.30}$$

Exploiting the properties of DCT matrices and the symmetry in the response, Yim [165] proposed an efficient computation in the $8 \times 8$ block DCT space.

### 3.4.2 Overlapping and Add Methods in the Transform Domain

In these methods, the contribution of an input block to the filtered response of neighboring blocks and also to itself is measured in isolation from others [107]. In the block DCT domain, this is computed in three steps, namely,

1. The block, $X$, is padded with zero neighbors and merged into a single larger block, say $X_l$.

2. Apply CMPs to the larger block for computing the filtered response in the transform domain.

3. Decompose the larger block into smaller ones, each providing the respective contribution to the neighboring blocks (including the central one).

Finally, the sum total of all such contributions from its neighbors, including its own response, provides the filtered response for the block. The first and third steps in the above involve multiplication of DCT blocks with block composition and decomposition matrices as discussed in the previous chapter. We use the same notation for denoting them in subsequent discussion. Accordingly, $A_{(L,N)}$ denotes the block composition matrix for merging $L$ adjacent $N$-point DCT blocks into a single $LN$ point DCT block.

For computing the filtered output of any general arbitrary response (that is, noncausal asymmetric FIR), only *two* of the CMPs as described in Table 3.1 are used. They correspond to the the first and last rows of the table. As the last one involves the DST, we need to convert DCT blocks into a DST block. We refer to this operation as *DST composition*. The DST composition matrix for transforming $L$ adjacent DCT blocks of length $N$ into a DST block of length $LN$ is denoted by $B_{(L,N)}$ and is given by the following expression.

$$B_{(L,N)} = S_{LN} \begin{bmatrix} C_N^T & 0_N & \cdots & 0_N \\ 0_N & C_N^T & \cdots & 0_N \\ \cdot & \cdot & \cdots & \cdot \\ 0_N & 0_N & \cdots & C_N^T \end{bmatrix}. \qquad (3.31)$$

As before, in the above equation, $S_{LN}$ is the $LN$-point type-II DST matrix, and $C_N$ is the $N$-point type-II DCT matrix. In subsequent subsections we present the properties and theorems related to filtering with different types of FIRs.

### 3.4.2.1  Filtering with Symmetric FIR

Let us consider the impulse response as $h(n), -L \le n \le L$. It is symmetric if $h(n) = h(-n)$. From the related CMP, as discussed earlier, it is sufficient to compute with its positive half $h(n), 0 \le n \le L$. Let it be denoted as $h^+(n)$. However, to make the operation equivalent to the linear convolution, we need to compute with an extended neighborhood. The extent of this neighborhood depends upon the length of the filter (that is, $2L+1$). So, given a block size $N$, the number of adjacent blocks to be merged is $\lceil \frac{L}{N} \rceil$. Let us restrict ourselves to the case $L \le N$. Hence, only two neighboring blocks are to be merged during the computation. We now state a theorem accounting for the contributions from neighboring blocks of the $i$th input block.

**Theorem 3.5** *Consider a matrix $U$ of size $3N \times 3N$ such that $U = A_{(3,N)}^T \mathbb{D}(\{\sqrt{6N}C_{3N}^I \mathbf{h}^+\}_0^{3N-1})A_{(3,N)}$ and its $N \times N$ submatrices $U_{ij}, 1 \le i, j \le 3$ are defined as follows:*

$$U = \begin{bmatrix} U_{11} & U_{12} & U_{13} \\ U_{21} & U_{22} & U_{23} \\ U_{31} & U_{32} & U_{33} \end{bmatrix}.$$

*Let $X_i$ be the $i$-th DCT block, and $Y_i$ be the DCT of the filtered output at the same location. Then, $Y_i$ is given by the following expression:*

$$Y_i = \begin{bmatrix} E_s & F_s & G_s \end{bmatrix} \begin{bmatrix} X_{i-1} \\ X_i \\ X_{i+1} \end{bmatrix}. \qquad (3.32)$$

*where $E_s = U_{32}$, $F_s = U_{22}$, and $G_s = U_{12}$.* □

**Proof:**

   Let us consider the contribution of a single block $X$ to its neighboring blocks. For this we compute the filtered response by composing a larger block surrounding $X$ with zeroes. This computation is based on the same *overlap and add* principle and is shown in the following:

$$Y = U \begin{bmatrix} 0_N \\ X \\ 0_N \end{bmatrix},$$

(from Eq. (2.94) and Eq. (2.110)). (3.33)

$$= \begin{bmatrix} U_{12}X \\ U_{22}X \\ U_{32}X \end{bmatrix}.$$

From Eq. (3.33) we determine the contributions of the $(i-1)$th and $(i+1)$th neighboring blocks toward the $i$-th block, and its response $Y_i$ is expressed in the form of Eq. (3.32). □

The matrices $E_s$, $F_s$, and $G_s$ are referred to as *filtering matrices*. From Theorem 3.5 we obtain their relationship with submatrices of $A_{3,N}$. Let us use the same notation for denoting the the submatrices of $A_{3,N}$ as $A_{i,j}, 0 \le i, j \le 2$, each of dimension $N \times N$. Similarly, we also denote diagonal submatrices of $\mathbb{D}(\{\sqrt{6N}C_{3N}^I \mathbf{h}^+\}_0^{3N-1})$ as $D_i = D_{i,i}, 0 \le i \le 2$. With these, the filtering matrices are expressed in the following form.

$$\begin{aligned} E_s &= A_{0,2}^T D_1 A_{0,1} + A_{1,2}^T D_2 A_{1,1} + A_{2,2}^T D_3 A_{2,1}, \\ F_s &= A_{0,1}^T D_1 A_{0,1} + A_{1,1}^T D_2 A_{1,1} + A_{2,1}^T D_3 A_{2,1}, \\ G_s &= A_{0,0}^T D_1 A_{0,1} + A_{1,0}^T D_2 A_{1,1} + A_{2,0}^T D_3 A_{2,1}. \end{aligned}$$ (3.34)

**Theorem 3.6** *Filtering matrices (as defined in Theorem 3.5) have the following properties.*

1. *$F_s$ is a symmetric matrix.*

2. *$F_s(k,l) = 0$ if $k+l$ is odd.*

3. *$E_s(k,l) = (-1)^{k+l} G_s(k,l)$.*

□

**Proof:**

For proving these properties, we take the help of Theorems 2.14 and 2.15 as shown below:

1. $U(= A_{(3,N)}^T \mathbb{D}(\{\sqrt{6N}C_{3N}^I \mathbf{h}^+\}_0^{3N-1}) A_{(3,N)}$ of Theorem 3.5) is symmetric as it is in the form of $UDU^{-1}$, where $D$ is a diagonal matrix. Hence, $F_s(= U_{1,1})$ is also symmetric. □

2. Let us consider a matrix $S_i = A_{i,1}^T D_i A_{i,1}$. Let the $t$th diagonal element (at $t$th row) of $D_i$ be denoted as $d(t)$. Hence, an element of $S_i$ is expressed

as follows:

$$S_i(k,l) = \sum_{t=0}^{N-1} A_{i,1}^T(k,t)d(t)A_{i,1}(t,l),$$

$$= \sum_{t=0}^{N-1} A_{i,1}(t,k)d(t)A_{i,1}(t,l). \tag{3.35}$$

Now, from Theorem 2.14, if $k$ and $l$ are of opposite polarities, either $A_{i,1}(t,k) = 0$ or $A_{i,1}(t,l) = 0$. This implies that $S(k,l)$ is zero if $k+l$ is odd. As $F_s = \sum_{i=0}^{2} S_i$, $F_s(k,l) = 0$ if $k+l$ is odd. $\qquad\square$

3. Let $S_{i,0} = A_{i,0}^T D_i A_{i,1}$ and $S_{i,2} = A_{i,2}^T D_i A_{i,1}$. From Theorem 2.15, $A_{i,0}(k,l) = (-1)^{k+l} A_{i,2}(k,l)$. Hence,

$$S_{i,0}(k,l) = \sum_{t=0}^{N-1} A_{i,0}^T(k,t)d(t)A_{i,1}(t,l),$$

$$= \sum_{t=0}^{N-1} A_{i,0}(t,k)d(t)A_{i,1}(t,l), \tag{3.36}$$

$$= \sum_{t=0}^{N-1} (-1)^{t+k} A_{i,2}(t,k)d(t)A_{i,1}(t,l).$$

Next we consider two cases separately, one for $l$ as even and the other for $l$ as odd.

For even $l$: In Eq. (3.36) all $A_{i,1}(t,l)$ for *odd* $t$ becomes zero (see Theorem 2.14). Hence,

$$S_{i,0}(k,l) = \sum_{m=0}^{\frac{N}{2}-1} (-1)^{2m+k}(A_{i,2}(2m,k)$$
$$d(2m)A_{i,1}(2m,l))$$
$$= (-1)^k \sum_{m=0}^{\frac{N}{2}-1} (A_{i,2}(2m,k) \tag{3.37}$$
$$d(2m)A_{i,1}(2m,l))$$
$$= (-1)^k S_{i,2}(k,l)$$

For odd $l$: In Eq. (3.36) all $A_{i,1}(t,l)$ for *even* $t$ becomes zero (see Theorem

2.14). Hence,

$$
\begin{aligned}
S_{i,0}(k,l) &= \sum_{m=0}^{\frac{N}{2}-1} (-1)^{2m+1+k}(A_{i,2}(2m+1,k) \\
&\qquad d(2m+1)A_{i,1}(2m+1,l)) \\
&= (-1)^{k+1} \sum_{m=0}^{\frac{N}{2}-1} (A_{i,2}(2m+1,k) \\
&\qquad d(2m+1)A_{i,1}(2m+1,l)) \\
&= (-1)^{k+1} S_{i,2}(k,l)
\end{aligned}
\tag{3.38}
$$

From Eqs. (3.37) and (3.38), it follows that

$$
S_{i,0}(k,l) = (-1)^{k+l} S_{i,2}(k,l)
\tag{3.39}
$$

As $E_s = \sum_{i=0}^{2} S_{i,0}$ and $G_s = \sum_{i=0}^{2} S_{i,2}$,

$E_s(k,l) = (-1)^{k+l} G_s(k,l)$.

$\square$

### 3.4.2.2 Filtering with Antisymmetric FIR

In this case, the approach is similar to the previous one. However, it applies a different CMP and models skewcircular convolution in its equivalent computation. A filter response $h(n)$ is antisymmetric when $h(n) = -h(-n)$. Note that, in this case, $h(0) = 0$. Hence, the complete specification of the filter can be obtained from the *strict positive half* of the response such as $h(n), 1 \le n \le L$. Let this portion of the response be denoted as $h^p(n)$. Here, the filtered response is computed in the DCT space by applying the last CMP of Table 3.1. This demands conversion of adjacent input DCT blocks into a DST block. The input–output relationship for filtering with antisymmetric FIR is expressed in the following theorem.

**Theorem 3.7** *Consider a matrix $V$ of size $3N \times 3N$ such that $V = A_{(3,N)}^T \mathbb{D}_{-1}(\{\sqrt{6N}S_{3N}^I \mathbf{h}^\mathbf{P}\}_1^{3N-1})B_{(3,N)}$ and its $N \times N$ submatrices $V_{ij}, 1 \le i, j \le 3$, are defined as follows:*

$$
V = \begin{bmatrix} V_{11} & V_{12} & V_{13} \\ V_{21} & V_{22} & V_{23} \\ V_{31} & V_{32} & V_{33} \end{bmatrix}.
$$

*Let $X_i$ be the $i$-th DCT block, and $Y_i$ be the DCT of the filtered output at the same location. Then, $Y_i$ is given by the following expression:*

$$
Y_i = -\begin{bmatrix} E_a & F_a & G_a \end{bmatrix} \begin{bmatrix} X_{i-1} \\ X_i \\ X_{i+1} \end{bmatrix},
\tag{3.40}
$$

*where $E_a = V_{32}$, $F_a = V_{22}$, and $G_a = V_{12}$.*

$\square$

Similar to the filtering matrices of symmetric FIR, $E_a$, $F_a$, and $G_a$ are found to have the following properties. The proof is of similar in nature and is omitted from the text.

**Theorem 3.8** *Filtering matrices (as defined in Theorem 3.7) have the following properties.*

1. $F_a(k,l) = 0$, *if* $k + l$ *is even.*

2. $E_a(k,l) = (-1)^{k+l+1} G_a(k,l)$.

$\square$

### 3.4.2.3    Filtering with an Arbitrary FIR

By using the above two methods, we compute filtered output for any arbitrary FIR in the block DCT space. This is achieved by expressing any arbitrary FIR $h(n)$ as a sum of symmetric response $h_s(n)$ and antisymmetric response $h_a(n)$ such that $h_s(n) = \frac{1}{2}(h(n) + h(-n))$, and $h_a(n) = \frac{1}{2}(h(n) - h(-n))$. Let us rename $h^+(n)$ as $h_s(n), 0 \le n \le L$, and $h^p(n)$ as $h_a(n), 1 \le n \le L$. Hence, by applying both the Theorems 3.5 and 3.7, we easily obtain the following input–output relationship.

**Theorem 3.9**

$$Y_i = \begin{bmatrix} E & F & G \end{bmatrix} \begin{bmatrix} X_{i-1} \\ X_i \\ X_{i+1} \end{bmatrix}, \qquad (3.41)$$

*where* $E = E_s - E_a$, $F = F_s - F_a$ *and* $G = G_s - G_a$. $\square$

For symmetric $h(n)$, $E_a = F_a = G_a = 0_N$. Similarly, for antisymmetric $h(n)$, $E_s = F_s = G_s = 0_N$. For a *causal* impulse response, $G = 0_N$, implying $G_s = G_a$. Again, from Theorems 3.6 and 3.8 in this case (for a causal response), $E = 2E_s = -2E_a$.

### 3.4.2.4    Efficient Computation

In each of the above cases, computation essentially involves multiplication and addition of matrices of dimensions $N \times N$. Instead of performing computation by direct multiplication and addition of filtering matrices, we adopt a more efficient policy by using modified filtering matrices as derived from the original ones:

$$\begin{aligned} P_s &= \frac{(E_s + G_s)}{2}, \\ Q_s &= \frac{(E_s - G_s)}{2}, \\ P_a &= \frac{(E_a + G_a)}{2}, \\ Q_a &= \frac{(E_a - G_a)}{2}. \end{aligned} \qquad (3.42)$$

The choice of these derived matrices is primarily motivated by the fact that they are relatively sparse, as can be observed from the properties of Theorems 3.6 and 3.8. The following theorem states the extent of sparsity of these matrices.

**Theorem 3.10** *Each of the matrices $F_s$, $P_s$, $Q_s$, $F_a$, $P_a$, and $Q_a$ has at least $\frac{N}{2}$ zeroes in each row or each column. Hence, the number of multiplications $(n_m)$ and additions $(n_a)$ required for multiplying one of these matrices with an input block of size $N$ are given as*

$$\begin{aligned} n_m &= \frac{N^2}{2}, \\ n_a &= (\frac{N}{2} - 1)N. \end{aligned}$$ (3.43)

$\square$

Using derived filtering matrices, efficient computation for both the *noncausal symmetric* and *antisymmetric* filtering operations are designed. As steps of these computations are similar for both cases, they are shown in Table 3.3. In this table, the corresponding filtering matrices are denoted by $P$, $Q$, and $F$, respectively. Their roles should be considered under the respective context of the type of responses. Instead of computing all individual contributions of neighbors at the same stage, for efficient handling of input data, contributions of a single block (say, the $i$th block) to its neighboring blocks are computed at the same time. They are subsequently buffered and pipelined in the process of accumulation for both the $(i-1)th$ and $(i+1)th$ blocks (see the last step of Table 3.3).

**Table 3.3**: Computational steps and associated complexities of filtering an input DCT block of size $N$ with symmetric or nonsymmetric responses

| Steps | $n_m$ | $n_a$ |
|---|---|---|
| $K = PX_i$ | $\frac{N^2}{2}$ | $N(\frac{N}{2} - 1)$ |
| $L = QX_i$ | $\frac{N^2}{2}$ | $N(\frac{N}{2} - 1)$ |
| $W_i = FX_i$ | $\frac{N^2}{2}$ | $N(\frac{N}{2} - 1)$ |
| $U_i = K + L$ | - | $N$ |
| $V_i = K - L$ | - | $N$ |
| $Y_i = U_{i+1} +$ | - | |
| $W_i + V_{i-1}$ | - | $2N$ |
| Total | $\frac{3}{2}N^2$ | $\frac{N}{2}(3N + 2)$ |

In the more general case of filtering with arbitrary noncausal and causal response, it is observed that the sparsity of derived matrices does not provide any additional advantage. In this case, it is required to employ direct matrix operations with original filter matrices such as $E$, $F$, and $G$. For noncausal cases, it takes $3N^2$ multiplications and $3N^2 - N$ additions for $N$ samples,

Table 3.4: Per sample computational cost of filtering in 1-D

| Impulse response type | $n_m$ | $n_a$ |
|---|---|---|
| Symmetric | $\frac{3N}{2}$ | $\frac{3N+2}{2}$ |
| Antisymmetric | $\frac{3N}{2}$ | $\frac{3N+2}{2}$ |
| Causal | $2N$ | $2N - 1$ |
| Noncausal arbitrary | $3N$ | $3N - 1$ |

whereas for causal filters the numbers are $2N^2$ and $2N^2 - N$, respectively. Computational complexities of different cases are summarized in Table 3.4.

## 3.5 Filtering 2-D Images

The methods discussed above for 1-D are easily extended to 2-D if the impulse response is separable. However, for nonseparable cases, we need to employ different strategies. We discuss below the filtering of images with both separable and nonseparable filters in the DCT domain.

### 3.5.1 Separable Filters

For separable filters, a finite impulse $h(m, n)$ can be written in the form of $h(m, n) = h_1(m)h_2(n), -L \leq m, n \leq L$. Let us represent the $(i, j)th$ input block as $\mathbf{x_{i,j}}$ in 2-D, and its corresponding type-II DCT as $\mathbf{X_{i,j}}$. The input-output theorems in 1-D can easily be extended to 2-D. We present below the extension of Theorem 3.9 as it is stated in a more generic framework.

**Theorem 3.11** *Let $h(m, n) = h_1(m)h_2(n), -L \leq m, n \leq L$ be the finite impulse response of a separable filter. Let us express each kernel function ($h_1(n)$ or $h_2(n)$) of the separable function as a sum of a pair of symmetric and antisymmetric functions as described below:*

$$
\begin{aligned}
h_{1s}(n) &= \tfrac{1}{2}(h_1(n) + h_1(-n)), \\
h_{1a}(n) &= \tfrac{1}{2}(h_1(n) - h_1(-n)), \\
h_{2s}(n) &= \tfrac{1}{2}(h_2(n) + h_2(-n)), \\
h_{2a}(n) &= \tfrac{1}{2}(h_2(n) - h_2(-n)).
\end{aligned}
\tag{3.44}
$$

*Let $E_{1s}$, $F_{1s}$, and $G_{1s}$ be the filter matrices corresponding to the response $h_{1s}(n)$ formed according to Theorem 3.5. Similarly, $E_{2s}$, $F_{2s}$, and $G_{2s}$ are formed from $h_{2s}(n)$. For the antisymmetric component $h_{1a}(n)$ ( $h_{2a}(n)$ ), these are $E_{1a}$ ($E_{2a}$), $F_{1a}$ ($F_{2a}$), and $G_{1a}$ ($G_{2a}$), respectively (from Theorem 3.7).*

Table 3.5: Per-pixel computational cost of filtering in 2-D

| Impulse response type | $n_m$ | $n_a$ |
|---|---|---|
| Symmetric | $3N$ | $3N + 2$ |
| Antisymmetric | $3N$ | $3N + 2$ |
| Causal | $4N$ | $4N - 2$ |
| Noncausal arbitrary | $6N$ | $6N - 2$ |

*Then, the DCT of the filtered output of $(i, j)$th block, $Y_{i,j}$, is given by the following expression:*

$$Y_{i,j} = \begin{bmatrix} E_1 & F_1 & G_1 \end{bmatrix} \begin{bmatrix} X_{i-1,j-1} & X_{i-1,j} & X_{i-1,j+1} \\ X_{i,j-1} & X_{i,j} & X_{i,j+1} \\ X_{i+1,j-1} & X_{i+1,j} & X_{i+1,j+1} \end{bmatrix} \begin{bmatrix} E_2^T \\ F_2^T \\ G_2^T \end{bmatrix}, \qquad (3.45)$$

*where $E_1 = E_{1s} - E_{1a}$, $F_1 = F_{1s} - F_{1a}$, and $G_1 = G_{1s} - G_{1a}$. Similarly, $E_2 = E_{2s} - E_{2a}$, $F_2 = F_{2s} - F_{2a}$, and $G_2 = G_{2s} - G_{2a}$.*

□

Due to the separability property of the impulse response, filtered outputs are computed in two stages. First, the computation is carried out along the vertical direction, and next, it is applied along the horizontal direction. Hence, for separable responses, the computation in Theorem 3.11 is expressed in the following equivalent form:

$$\begin{aligned} Z_{i,j} &= \begin{bmatrix} E_1 & F_1 & G_1 \end{bmatrix} \begin{bmatrix} X_{i-1,j} \\ X_{i,j} \\ X_{i+1,j} \end{bmatrix}, \\ Y_{i,j} &= \begin{bmatrix} Z_{i,j-1} & Z_{i,j} & Z_{i,j+1} \end{bmatrix} \begin{bmatrix} E_2^T \\ F_2^T \\ G_2^T \end{bmatrix}. \end{aligned} \qquad (3.46)$$

For each stage of computation we adopt similar computation strategies of 1-D as charted in Table 3.3. Thus, per-pixel computational costs for different types of filter responses are given in Table 3.5. Similar strategies are also adopted for other methods [79, 165] in 1-D.

### 3.5.1.1 Sparse Computation

In the block DCT space, we may opt for computing with a sparse DCT block. As in most cases, higher frequency components are of smaller magnitudes, they may not be considered during computation, and only first $\frac{N}{2} \times \frac{N}{2}$ low-frequency coefficients are used for this purpose. This may be applicable to a few selected blocks, even at times for all the input blocks. This type of computation is termed here as *sparse computation*. In the *overlapping and add*

Table 3.6: Per-pixel computational cost of filtering in 2-D using the ASC

| Impulse response type | $n_m$ | $n_a$ |
|---|---|---|
| Symmetric | $\frac{3N}{8}$ | $\frac{3N+20}{16}$ |
| Antisymmetric | $\frac{3N}{8}$ | $\frac{3N+20}{16}$ |
| Causal | $\frac{N}{2}$ | $\frac{N-1}{2}$ |
| Noncausal arbitrary | $\frac{3N}{4}$ | $\frac{3N-2}{4}$ |

Table 3.7: Per-pixel computational cost of filtering in 2-D using the SNC

| Impulse response type | $n_m$ | $n_a$ |
|---|---|---|
| Symmetric | $\frac{5N}{4}$ | $\frac{5N}{4} - 1$ |
| Antisymmetric | $\frac{5N}{4}$ | $\frac{5N}{4} - 1$ |
| Causal | $\frac{9N}{4}$ | $\frac{9N+6}{4}$ |
| Noncausal arbitrary | $\frac{5N}{2}$ | $\frac{5N+2}{2}$ |

method discussed here, we consider two variations of this sparse computation. First, all the input DCT blocks are taken as sparse. This is referred to as *all sparse computation* (ASC). In the second variation, the neighboring blocks are considered as sparse. In the latter case, all the DCT coefficients of the central block are used in the computation. This variation is called *sparse neighbor computation* (SNC). We also refer to the *nonsparse computation*, that is, filtering with all the coefficients of every DCT block, as *full block computation* (FBC). Per-pixel computational costs of these two techniques are given in Tables 3.6 and 3.7, respectively.

### 3.5.1.2 Computation through Spatial Domain

As explained in Chapter 1, the cost of computation through the spatial domain should include those of inverse and forward transformation of the DCT. In this regard, the computational costs of these transforms are accounted for by the efficient technique as reported in [82]. According to this method, the required number of multiplications ($n_m$) and additions ($n_a$) for 2-D $N \times N$ DCT transform are given below.

$$
\begin{aligned}
n_m &= \frac{N^2}{2} log_2 N, \\
n_a &= \frac{3N^2}{2} log_2 N - N^2 + N.
\end{aligned}
\tag{3.47}
$$

**Table 3.8**: Per-pixel computational cost of filtering through convolution in the spatial domain

| Impulse response type | $n_m$ | $n_a$ |
|---|---|---|
| Symmetric | $2N + 2 + log_2 N$ | $2N + 3log_2 N - 2 + \frac{2}{N}$ |
| Antisymmetric | $2N + log_2 N$ | $2N + 3log_2 N - 4 + \frac{2}{N}$ |
| Causal | $2N + 2 + log_2 N$ | $2N + 3log_2 N - 2 + \frac{2}{N}$ |
| Noncausal arbitrary | $4N + 2 + log_2 N$ | $4N + 3log_2 N - 2 + \frac{2}{N}$ |

Let a separable filter of nonuniform kernel $h(m, n) = h_1(m)h_2(n), -N \leq m, n \leq N$, be applied to an image. In the spatial domain, the convolution operation is efficiently implemented by a sequence of three operations, namely, *shift*, *multiplication*, and *addition*. These are carried out for each nonzero $h_1(m)$ and $h_2(n)$ in two stages. In the first stage, the shifts are along vertical directions and in the second stage, these are applied in horizontal directions on the data obtained from the first stage. Following this implementation per-pixel computational complexities of these filters are shown in Table 3.8.

From Tables 3.5 and 3.8, it is observed that, for small block sizes, symmetric or antisymmetric filtering in the block DCT domain perform better than the spatial domain technique. However, the latter is superior to DCT filtering for more general types of impulse responses. However, the ASC or the SNC technique (see Tables 3.6 and 3.7) offers significant savings in the computation. It is of interest to know how the quality of the results suffers due to them. This is discussed in the next subsection with typical examples of filtering.

### 3.5.1.3  Quality of Filtered Images with Sparse Computation

The quality of filtered output is judged with reference to the output obtained by the linear convolution operation in the spatial domain. In this regard, the *peak signal-to-noise ratio* (PSNR) is an appropriate measure. Moreover, to observe the blocking and blurring artifacts due to the introduction of discontinuities at boundaries of $8 \times 8$ blocks, the *JPEG quality metric* (JPQM) is used. As discussed previously (see Section 1.7.2.2), the higher the value of JPQM, the better is the quality of the image, and for an image with good visual quality, the JPQM value should be close to 10 or higher.

The results presented here are obtained by applying separable filters of uniform kernels in the form of $g(m, n) = h(m)h(n)$. Typically, we have chosen a set of kernels as given in Tables 3.9, 3.10, 3.11, and 3.12 for different types of filter response. For each case, the FBC, the ASC, and the SNC algorithms are used. The PSNR and JPQM values obtained on application of these filters over a typical image *Lena* (of size $256 \times 256$) are shown in Table 3.13.

From Table 3.13, it is observed that the PSNR values obtained from the

Table 3.9: A noncausal symmetric FIR

| $n$ | -3 | -2 | -1 | 0 | 1 | 2 | 3 |
|---|---|---|---|---|---|---|---|
| $h(n)$ | $\frac{1}{12}$ | $-\frac{1}{6}$ | $\frac{1}{3}$ | $\frac{1}{2}$ | $\frac{1}{3}$ | $-\frac{1}{6}$ | $\frac{1}{12}$ |

Table 3.10: A noncausal antisymmetric FIR

| $n$ | -3 | -2 | -1 | 0 | 1 | 2 | 3 |
|---|---|---|---|---|---|---|---|
| $h(n)$ | $-\frac{1}{12}$ | $\frac{1}{6}$ | $-\frac{1}{3}$ | 0 | $\frac{1}{3}$ | $-\frac{1}{6}$ | $\frac{1}{12}$ |

FBC are very high ($> 300$ dB), implying that the FBC in the block DCT space is an equivalent technique of the linear convolution in the spatial domain. We also observe that all the JPQM values obtained from these techniques are significantly high. However, these values vary depending on filter response and input data. We also compare PSNR values obtained from the ASC and the SNC. As the PSNR values obtained from the ASC are significantly less than those obtained from the SNC, the latter technique is better suited for different applications. Though its computational cost is higher than that of the ASC, it is substantially low compared to the FBC, and computationally more efficient than filtering through spatial domain operations (see Tables 3.8 and 3.7).

### 3.5.2 Nonseparable Filters

Due to the nonseparability of the response, it is not possible to translate the convolution–multiplication property in 2-D into a linear form. Rather, in this case, we need to perform point-wise multiplication between the coefficients in the transform space. We restrict our discussion to filtering with *symmetric* nonseparable finite impulse responses. Let $h(m,n), -L \leq m,n \leq L$ be denoted as the impulse response. As it is symmetric, $h(m,n) = h(-m,-n)$. Hence, the specification in any quadrant in the 2-D space is sufficient enough to describe the complete response. Let us denote the response in the first quadrant as $h^+(m,n) = h(m,n), 0 \leq m,n \leq L$. For block filtering, we follow the same strategy of merging the neighboring blocks of an input block. Then, the convolution–multiplication property of 2-D transforms is applied by pointwise multiplication of type-I DCT coefficients of the response and type-II DCT coefficients of the merged DCT block. This property is an extended form of what is shown in the first row of Table 3.1. Finally, block decomposition is performed on the larger filtered block, and the central target block is saved in the resulting output. The above computational steps are summarized in the theorem given below.

Table 3.11: A noncausal FIR

| $n$ | -3 | -2 | -1 | 0 | 1 | 2 | 3 |
|------|------|------|------|------|------|------|------|
| $h(n)$ | $\frac{1}{12}$ | $-\frac{1}{6}$ | $\frac{1}{3}$ | $\frac{1}{2}$ | $\frac{1}{4}$ | $-\frac{1}{12}$ | $\frac{1}{12}$ |

Table 3.12: A causal FIR

| $n$ | 0 | 1 | 2 | 3 |
|------|------|------|------|------|
| $h(n)$ | $\frac{1}{2}$ | $\frac{1}{4}$ | $\frac{1}{8}$ | $\frac{1}{8}$ |

**Theorem 3.12** *Let* $h(m,n), -L \leq m, n \leq L$ *be the symmetric finite impulse response, and its response in the first quadrant is denoted as* $h^+(m,n) = h(m,n), 0 \leq m, n \leq L$. *Let* $X_{i,j}$ *be the* $(i,j)$th $N \times N$ *input DCT block.*

*Then, the DCT of the filtered output of* $(i,j)$th *block,* $Y_{i,j}$, *is given by the following expression:*

$$
Y_{i,j} =
$$
$$
PA^T_{(3,N)} \left( A_{(3,N)} \begin{bmatrix} X_{i-1,j-1} & X_{i-1,j} & X_{i-1,j+1} \\ X_{i,j-1} & X_{i,j} & X_{i,j+1} \\ X_{i+1,j-1} & X_{i+1,j} & X_{i+1,j+1} \end{bmatrix} A^T_{(3,N)} \otimes H \right) A_{(3,N)} P^T,
$$
$$
(3.48)
$$

*where* $A_{(3,N)}$ *is the block composition matrix as discussed earlier.* $H$ *is the type-I DCT of the response in the first quadrant. In our notation, a matrix formed by the elements* $h(m,n), a \leq m \leq b, c \leq n \leq d$ *is represented as* $\{h(m,n)\}_{a \leq m \leq b, c \leq n \leq d}$. *Thus* $H$ *is given by*

$$
H = 6N\{C_{1e}\left(\{h^+(m,n)\}_{0 \leq m,n \leq 3N}\right)(k,l)\}_{0 \leq k,l \leq 3N-1}.
$$

**Table 3.13**: PSNR (in DB) and JPQM values of filtered output using proposed techniques

| Type | $h(n)$ (from Tables) | FBC | | ASC | | SNC | |
|------|------|------|------|------|------|------|------|
| | | PSNR | JPQM | PSNR | JPQM | PSNR | JPQM |
| Symmetric | 3.9 | 302.21 | 10.89 | 36.58 | 11.41 | 44.25 | 12.01 |
| Antisymmetric | 3.10 | 329.05 | 14.62 | 45.91 | 19.19 | 50.44 | 16.99 |
| Causal | 3.12 | 304.43 | 10.82 | 41.60 | 9.43 | 53.89 | 10.67 |
| Noncausal arbitrary | 3.11 | 304.26 | 10.93 | 37.66 | 11.08 | 45.92 | 11.82 |

*P is the selection matrix as given below.*

$$P = \begin{bmatrix} 0_N & I_N & 0_N \end{bmatrix}.$$

$0_N$ *and* $I_N$ *are* $N \times N$ *zero and identity matrices, respectively. The operator* $\otimes$ *denotes the elementwise multiplication.*  □

The above technique [104] follows the *overlap and save* approach.

## 3.6    Application of Filtering

Filtering is used in different applications of image and video processing. In subsequent chapters, we find its various applications in the development of algorithms for image and video processing in the compressed domain. In this section, let us consider two typical cases, namely, removal of blocking artifacts and sharpening of images.

### 3.6.1    Removal of Blocking Artifacts

Reconstruction from highly compressed images suffers greatly from blocking artifacts. In Figure 3.2(b), we demonstrate an image compressed by JPEG compression algorithm at quality factor 10. We can see the annoying presence of blocking artifacts in the image. This is also implied by a very low JPQM value (3.88) of the image. Using low-pass filtering, we mask these blocking artifacts substantially. Such a processed image with a Gaussian filter of standard deviation 2.0 is presented in Figure 3.2(c). The blocking artifacts are significantly reduced in this image, which is also highlighted by a remarkable improvement of its JPQM value (10.74).

### 3.6.2    Image Sharpening

To sharpen an image, we add a fraction of the high-pass filtered output with it. As DCT is a linear transform, we can simply add the difference from its low-pass filtered DCT blocks, scaled by a weight, to itself. Let $B_f$ be the low-pass filtered block in the transform domain. Let $B$ be its corresponding original block. Hence, the sharpened block in the transform domain is computed as follows:

$$B_s = B + \lambda(B - B_f). \tag{3.49}$$

In Eq. (3.49), $\lambda(> 0)$ is the parameter controlling the amount of image sharpening. Figure 3.3 shows a sharpened images for $\lambda = 0.9$.

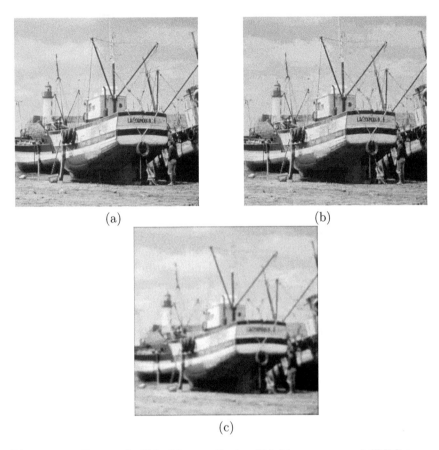

(a) (b)

(c)

**Figure 3.2**: Removal of blocking artifacts of highly compressed JPEG images using Gaussian filtering in the compressed domain:(a) original image, and (b) compressed at quality factor 10 (JPQM = 3.88), and (c) filtered image (JPQM = 10.74).

**Figure 3.3**: Image sharpening using Gaussian filtering with $\lambda = 0.9$ and $\sigma = 2.0$ (JPQM = 9.87).

## 3.7　Summary

Filtering of images in the block DCT space can be performed by the application of suitable convolution–multiplication properties of trigonometric transforms or by using the linear and distributive property of the DCT to the convolution operation of the spatial domain. However, we need to take care of discontinuities introduced due to block boundaries in the block DCT space. To obviate this problem, we adopt the strategy of block composition and decomposition. There are other approaches that consider a larger neighborhood in their computational model for the same purpose. The advantage of working with separable filter response is that it can trivially extend the linear computation model (involving a CMP) of 1-D to 2-D. However, for a nonseparable filter response, we need to perform the nonlinear operation of point-wise multiplication of the transform coefficients of the response and the input image block.

# Chapter 4

## Color Processing

With the rapid advancement in imaging technology, color images are more easily available and used in different applications than ordinary black and white or gray scale images. Increasing use of color has led to formulation of new challenges and problems in image processing. For example, various image capturing and display devices have their own limitations on data acquisition and rendition. Hence, different techniques and algorithms have been advanced to overcome their limitations and to preserve color information as images are exchanged between devices. Mostly, these algorithms are developed in the spatial domain. However, a few techniques have also been reported for processing color images in the compressed domain, in particular in the block discrete cosine transforms (DCT) domain. In this chapter we discuss some of the techniques and problems of color processing in the block DCT space. First, we briefly introduce the fundamentals related to color processing, which is followed by a discussion on different approaches in the compressed domain.

## 4.1  Color Representation

Newton's pioneering experiments on sunlight revealed the fact that the color of light depends upon its spectral components. Later, Grassmann [54] and Maxwell [97] laid the mathematical foundation of representation of a color factored into *three independent variables*. Grassmann stated the law of color matching [159], and Maxwell demonstrated that any additive color mixture could be matched by appropriate amount of three primary stimuli [98]. This is referred to as *trichromatic generalization or trichromacy*. For a long time, psychophysical experiments [159] were carried out to determine *color matching functions* (CMFs), which form a set of three functions providing numerical specification of color in terms of three coordinates or *tristimulus* values. Hence, it is the presence of different wavelengths of optical light and their relative strength that determine the sensation of color. This characteristics of a light source is expressed in terms of *spectral power distribution* (SPD). According to our perception, we attribute these three factors of color to the following.

1. **Brightness**: Brightness is the attribute of a visual sensation by which an area appears to emit light.

2. **Hue**: Hue is the attribute related to our perception of distinct colors such as red, yellow, green, blue, etc. If there is a shift in the dominant wavelength of the SPD, the hue of the associated color also shifts.

3. **Saturation**: Saturation indicates the purity of the perceived hue. It measures how an area is colorful in proportion to its brightness. It varies from neutral gray to saturated colors. The more the SPD is concentrated at a wavelength, the more saturated is the associated color. Again, for desaturating a color, we simply add white light that contains power at all wavelengths.

Out of the above three, brightness is the feature of the *achromatic* component of the color. Hue and saturation form the chromatic components of color representation.

## 4.2 Color Space

In an optical color camera, we have three sensors, each operating at different zones of the optical wavelengths, namely, with small wavelengths (blue zone), mid wavelength range (green zone), and large wavelengths (red zone). Let $r(\lambda)$, $g(\lambda)$, and $b(\lambda)$ be relative spectral responses of sensors corresponding to red, green, and blue zones of the optical spectrum. We also consider $E(\lambda)$ to be the SPD of the light source whose rays are reflected from an object surface point $X$ and projected on $x$ in the image plane. Let $\rho(\lambda)$ represent the surface reflectance spectrum at $X$. In a simplified model, we compute the output of an individual sensor by accumulating spectral responses over the range of wavelength $\lambda$ on which it is active. Assuming all the reflecting bodies as ideal 2-D flat Lambertian surfaces, the brightness values $R$, $G$, and $B$ for respective sensors at an image coordinate $x$ are expressed in the following form:

$$\begin{aligned} R &= \int_\lambda E(\lambda)\rho(\lambda)r(\lambda)d\lambda, \\ G &= \int_\lambda E(\lambda)\rho(\lambda)g(\lambda)d\lambda, \\ B &= \int_\lambda E(\lambda)\rho(\lambda)b(\lambda)d\lambda. \end{aligned} \tag{4.1}$$

### 4.2.1 RGB Color Space

In Eq. (4.1), the color of the pixel at $x$ in the image plane is represented by the vector $(R, G, B)$, whose each field denotes the amount of *red*, *green*, and *blue* to be mixed for producing the color according to our perception. The display devices also work on the same principle of mixing these three *primary colors* in the same proportion. The color space representing a color in this way is called the RGB color space. It may be noted that not all colors are reproducible by adding three primary colors. Some of them require subtraction of a primary color from the mixture of the other two. The major problem of representing a color in the RGB space is that it does not provide the perceptual factors of color representation, in particular, hue and saturation. That is why different other color spaces are proposed and used in various applications of color processing.

### 4.2.2 CIE XYZ Color Space

The International Commission on Illumination, better known as CIE (Commission Internationale d'Eclairage), is the primary organization that drives the standardization of color metrics and terminologies. In 1931, CIE defined its first colorimetry standard by providing two different but equivalent sets of CMFs. The first set of CMFs is called the CIE *red–green–blue* (RGB) CMFs, which are based on monochromatic primaries at wavelengths of 700 nm, 546.1 nm, and 435.8 nm, respectively [29]. The second set of CMFs, known as the

CIE XYZ CMFs, are defined in terms of a linear transformation of the CIE RGB CMFs [71]. The choice of this transformation was made in such a way that the CMF of the $Y$ component represents the *luminous efficiency function* [159]. The other consideration was to yield equal tristimulus values for the equienergy spectrum through normalization of the three CMFs. Moreover, the transformation takes care of the fact that all the CMFs become nonnegative. This makes the CIE primaries physically nonrealizable. The $Y$ tristimulus value is usually called the *luminance* and correlates with the perceived brightness of the radiant spectrum. The linear transformation from RGB to XYZ color space is given below.

$$
\begin{bmatrix} X \\ Y \\ Z \end{bmatrix} = \begin{bmatrix} 0.49 & 0.31 & 0.20 \\ 0.177 & 0.813 & 0.011 \\ 0.000 & 0.010 & 0.990 \end{bmatrix} \begin{bmatrix} R \\ G \\ B \end{bmatrix}. \tag{4.2}
$$

### 4.2.3   CIE Chromaticity Coordinates

Though a color is represented in a 3-D space, its chromatic information are represented in a 2-D space. One simple way to achieve this representation is to normalize the vector by its brightness value or the sum of the stimuli. This makes one of the coordinates redundant. It is obtained from the other two, as the three chromaticity coordinates sum up to unity. Hence, colors are represented using only two chromaticity coordinates. The plot of these points in the normalized 2-D space is known as the *chromaticity diagram*. The most commonly used chromaticity diagram is the CIE *xy chromaticity diagram*. The CIE xyz chromaticity coordinates are obtained from the $X, Y,$ *and* $Z$ tristimulus values in CIE XYZ space as given below:

$$
x = \frac{X}{X + Y + Z}, \tag{4.3}
$$

$$
y = \frac{Y}{X + Y + Z}, \tag{4.4}
$$

$$
z = \frac{Z}{X + Y + Z}. \tag{4.5}
$$

In Figure 4.1, the plot of the curve corresponding to the visible monochromatic spectra on the CIE xy chromaticity diagram is shown. This shark-fin-shaped curve, along which the wavelength (in nm) is indicated, is called the *spectrum locus*. In the figure, a few typical points corresponding to optical wavelengths are also marked. All the physically realizable colors lie in the region inside the closed curve formed by the spectrum locus, and the line joining its two extremes of 360 nm and 830 nm is known as the *purple line*.

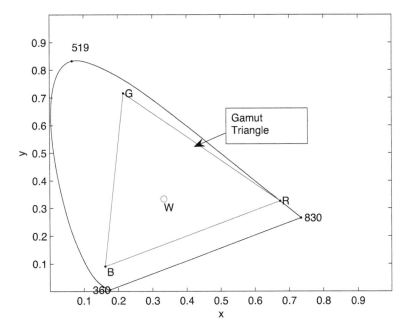

Figure 4.1: CIE chromaticity diagram

However, given the chormaticity coordinates of the three primary colors, the reproducible colors from their additive mixture lie within the convex hull or the triangle formed by them. This triangle is referred to as the *gamut triangle* (as shown in the figure), and the set of reproducible colors within it is called the *color gamut*. The coordinate point $(0.33, 0.33)$ is the achromatic point in the chromaticity chart. This point is known as the *reference white* point (marked as 'W' in the figure).

## 4.2.4   YCbCr Color Space

It was already discussed in Chapter 1 (see Section 1.4.1) that the color space used in the JPEG compression is the YCbCr color space. The transformation from the RGB color space is shown below:

$$\begin{bmatrix} Y \\ Cb \\ Cr \end{bmatrix} = \begin{bmatrix} 0.256 & 0.502 & 0.098 \\ -0.148 & -0.290 & 0.438 \\ 0.438 & -0.366 & -0.071 \end{bmatrix} \begin{bmatrix} R \\ G \\ B \end{bmatrix} + \begin{bmatrix} 0 \\ 128 \\ 128 \end{bmatrix}. \qquad (4.6)$$

In this space, $Y$ represents the *luminance component*, while $Cb$ and $Cr$ represent the chromatic parts. The above transformation is not linear. However, if $Cb$ and $Cr$ are translated by $-128$, it becomes linear. In that case, the corresponding color space is also known as YUV color space.

## 4.3    Processing Colors in the Compressed Domain

The foremost advantage of processing colors in the DCT space is that the chromatic components of the data are separated from its luminance component. This leads to two different approaches on color processing in the compressed domain. In one approach, it is the luminance component (Y) that is subjected to the processing, keeping the Cb and Cr unaltered. For example, in different color enhancement algorithms [1, 84, 141], computation is carried out in the luminance component only. This makes substantial reduction in the computation. However, this operation does not ensure the preservation of colors, and in some techniques [105], the chromatic components are also adjusted accordingly. In the other approach, processing is restricted to chromatic components only, keeping the Y component unaltered. For example, we may implement chromatic shift [39] for correcting colors under biased illumination of a scene.

The other advantage of DCT space is the spectral factorization of the data. This fact is exploited by treating DC and AC coefficients of each block differently. Moreover, the set of DC coefficients provides a low-resolution representation of the image. Using this set we estimate different statistics [109, 113] of the data and use them in the computation. For example, in [113], color palettes are recovered from DC coefficients of the compressed image. Similarly, in [109], the color constancy problem is solved by observing the statistical distribution of DC coefficients of chrominance components.

In this chapter, we discuss a few representative problems of color processing using one of the above approaches. In our discussion we consider the following operations for color image processing.

1. Color saturation and desaturation

2. Color constancy

3. Color enhancement

## 4.4    Color Saturation and Desaturation

How to enhance colors through the process of saturation and desaturation in the CIE xy coordinate space is discussed in [90]. The process is demonstrated in Figure 4.2. In this space the reference white ($W$) lies at $(\frac{1}{3}, \frac{1}{3})$. Any radial straight line coming out of this point is considered to have chromaticity points with a constant hue value. On the other hand, saturation of the color increases with the increasing radial distance from $W$. In reality, constant hue lines in xy space are not straight lines. For simplifying the computation, these curves are approximated by a straight line joining the point (say $p$ in Figure 4.2) and the reference white point ($W$). With this simplified representation, the maximally saturated point in the CIE space is defined as the intersection of $\overrightarrow{Wp}$ with an edge of the gamut triangle formed by the chromaticity points of the three primary colors in this space. This point is shown as $s$ in Figure 4.2. In the next stage, desaturation is carried out by radially translating back the saturated point (say, at $d$ in Figure 4.2) according to the *center of gravity law for color mixture* [61]. As points very near to $W$ are more whitish (less saturated), they are kept aside from this computation. These are determined by computing the distance from $W$, and points lying at a distance less than a threshold are not processed as above. In Figure 4.2, the interior of the circle around $W$ defines this region of unprocessed color points.

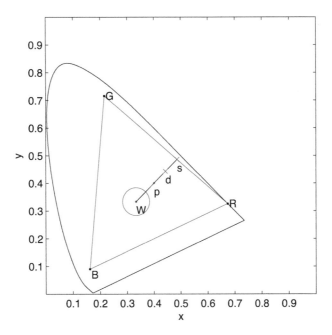

Figure 4.2: Saturation and desaturation operation in xy chromaticity space.

### 4.4.1 Normalized YCbCr Color Space

In the compressed domain, the foregoing computation becomes more efficient if it is carried out directly in the YCbCr color space. In that case, the computational overhead of conversion of color spaces gets eliminated. To this end, a normalized color space from the YCbCr is defined here. This space is referred to as the normalized $CbCr$ color space (nCbCr).

$$nCb = \frac{Cb - 128}{a.Y + b.Cb + c.Cr} \quad \text{and} \quad nCr = \frac{Cr - 128}{a.Y + b.Cb + c.Cr}. \tag{4.7}$$

where $a$, $b$, and $c$ are constants. From Eqs. (4.2) and (4.6), their values are obtained as 3.51, 1.99, and 0.14, respectively. The relationship between the CIE xy space and nCbCr space is given below:

$$\begin{bmatrix} nCb \\ nCr \end{bmatrix} = \begin{bmatrix} -0.6823 & -0.7724 \\ 1.532 & -0.6047 \end{bmatrix} \begin{bmatrix} x \\ y \end{bmatrix} + \begin{bmatrix} 0.4849 \\ -0.3091 \end{bmatrix}. \tag{4.8}$$

Similarly, a point $((x, y))$ in the xy space is obtained from the $(nCb, nCr)$ as follows:

$$\begin{bmatrix} x \\ y \end{bmatrix} = \begin{bmatrix} -0.3789 & 0.484 \\ -0.96 & -0.4275 \end{bmatrix} \begin{bmatrix} nCb \\ nCr \end{bmatrix} + \begin{bmatrix} \frac{1}{3} \\ \frac{1}{3} \end{bmatrix}. \tag{4.9}$$

In Figure 4.3 the chromaticity points of Figure 4.2 are shown in the nCbCr color space. We observe that the origin $((0, 0))$ of the space becomes the reference white point. This simplifies the computation of saturation and desaturation points. Moreover, the characterization of hue preserving transformation in the YCbCr space could be conveniently expressed in the following form.

**Theorem 4.1** *Two points in the $YCbCr$ space with color co-ordinates $(Y_1, Cb_1, Cr_1)$ and $(Y_2, Cb_2, Cr_2)$ have the same hue if $\frac{Cb_1 - 128}{Cr_1 - 128} = \frac{Cb_2 - 128}{Cr_2 - 128}$.*

□

From the above theorem, we observe that colors of same hue lie along the radial straight line emanating from the point $(128, 128)$ in the CbCr space.

### 4.4.2 Maximum Saturation

The adaptation of the method discussed in [90] for maximally saturating colors is described in the following algorithm.

*Algorithm Max_Saturate_Color* (MaxSat)

*Input:* A color point in YCbCr space denoted as $P = (Y, Cb, Cr)$, a distance threshold $\tau$. $\theta_1$, $\theta_2$

*Output:* Maximally saturated point with the same hue and luminance, say, $S = (Y, Cb_s, Cr_s)$.

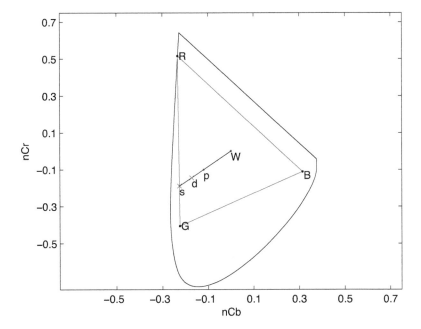

Figure 4.3: Saturation–desaturation in nCbCr space.

1. Convert $S$ into the coordinates in nCbCr space, say, $p = (nCb, nCr)$, using Eq. (4.7). Let the normalizing factor in Eq. (4.7) be represented as $F = a.Y + b.Cb + c.Cr$.

2. If $((\| p \| > \tau)$ and $(\theta_1 \leq tan^{-1}(\frac{nCr}{nCb}) \leq \theta_2))$, perform the following operations.

   2a. Compute the intersection between $\overrightarrow{Wp}$ and the gamut triangle as $s = (nCb_s, nCr_s)$.

   2b. Compute the maximally saturated chroma components as follows:

   $$\begin{bmatrix} Cb_s \\ Cr_s \end{bmatrix} = a.Y \begin{bmatrix} 1 - b.nCb & -c.nCb \\ -b.nCr & 1 - c.nCr \end{bmatrix}^{-1} \begin{bmatrix} nCb \\ nCr \end{bmatrix} + \begin{bmatrix} 128 \\ 128 \end{bmatrix}. \tag{4.10}$$

End *Algorithm Max_Saturate_Color*

In the above algorithm, the step 2b is required to preserve the RGB color vector in the saturated image. However, even without color preservation, the technique provides comparable results. Hence, the corresponding computation is approximated by multiplying $nCb_s$ and $nCr_s$ with $F$. This makes the computation faster without any noticeable deterioration of the result. In Figure 4.4 (**see color insert**), the maximally saturated image of the input image 'image22' [1] is shown. In this example, $\tau$ is kept at 0.05, whereas values of $\theta_1$ and $\theta_2$ are taken as $\frac{\pi}{2}$ and $2\pi$, respectively. In Figure 4.5 (a) and (b)(**See color insert**), chromaticity points for the original image and the maximally saturated image, respectively, are plotted in the nCbCr space. It is observed that many points are clustered along the edges of the gamut triangle in the latter plot, while points in the vicinity of the reference white ($W$) remain unchanged.

### 4.4.3    Desaturation of Colors

The computation of desaturation of colors also gets simplified compared to the methods discussed in [90], which uses the law of center of gravity for color mixing [61] for moving the chromaticity point of maximum saturation toward the reference white. In this computational model, the mass at the reference white is taken as the fraction of average luminance $Y_{avg}$ of the image. Let the corresponding fraction be denoted as $w$ ($w \in [0, 1]$). The luminance value at the saturation point (say, $Y$) is taken as its corresponding mass. Hence, in the nCbCr space, the chromaticity point (say, $(nCb_d, nCr_d)$) at the desaturation point (denoted as the point $d$ in Figure 4.3) is computed as follows.

$$\begin{bmatrix} nCb_d \\ nCr_d \end{bmatrix} = \frac{Y}{Y + w.Y_{avg}} \begin{bmatrix} nCb_s \\ nCr_s \end{bmatrix}. \tag{4.11}$$

Further, the luminance value in the desaturation point is taken as $Y + w.Y_{avg}$. This method is referred to in this work as *SatDesat*. In Figure 4.6 (**see color insert**), the desaturated image from the maximally saturated one is shown. In this case, the value of $w$ is empirically chosen as 0.2.

### 4.4.4    Computation in the Block DCT Space

The above algorithms in the YCbCr color space are developed for applying them directly in the block DCT space. Moreover, the computation reduces further if only the DC coefficients of individual blocks are processed. In Figure 4.7 (**see color insert**), the resulting images obtained from the two algorithms *MaxSat* and *SatDesat* are shown. Corresponding techniques in the DCT domain are referred to here as *MaxSat-DCT* and *SatDesat-DCT*. The resulting images are visually comparable to those obtained by the techniques in the spatial domain itself. However, due to the computation in the block DCT space, processed images suffer from blocking artifacts. In Table 4.1, the *JPEG quality metric* (JPQM) values are shown for the images in Figure 4.7, and they are found to be above 8. For observing the extent of color enhance-

---

[1]Obtained from the website *http://dragon.larc.nasa.govt/retinex/pao/news*

ment, *color enhancement factors* (CEF) (see Section 1.7.2.3 of Chapter 1) are also tabulated against different techniques. In addition to them, to observe how saturated images retain the distinguishing features and descriptions in an image, the *structural similarity index* (SSIM) [152] (see Section 1.7.2.1 of Chapter 1) between two images is computed. For a color image, the SSIM for all the three components $Y$, $Cb$, and $Cr$ are computed separately, and they are named Y-SSIM, Cb-SSIM, and Cr-SSIM, respectively. From the table, we observe that the performance in the block DCT domain is comparable with that of spatial domain techniques.

Table 4.1: Performance of different algorithms on the image *Bridge*

| Method | Y-SSIM | Cb-SSIM | Cr-SSIM | CEF | JPQM |
|---|---|---|---|---|---|
| MaxSat | 0.97 | 0.99 | 0.85 | 1.30 | — |
| SatDesat | 0.91 | 0.95 | 0.92 | 1.34 | — |
| MaxSat-DCT | 0.99 | 1.00 | 0.89 | 1.21 | 8.64 |
| SatDesat-DCT | 0.95 | 0.97 | 0.95 | 1.27 | 8.37 |

## 4.4.5 Computational Cost

Computation in the nCbCr space has an advantage while processing compressed images. As the inverse DCT of an image belongs to the YCbCr color space, the cost due to color space conversion is saved in all the algorithms suggested here. On the contrary, adoption of the existing algorithm [90] and its variation [114] requires this conversion.

### 4.4.5.1 MaxSat

In this algorithm, for processing each pixel, we require to (i) compute its chromaticity coordinates in the nCbCr space (with $5M + 2A$ operations), (ii) obtain the intersection with the gamut triangle (with $5M + 6A$ operations), and (iii) obtain $Cb$ and $Cr$ from the normalized coordinate of the saturation point (with $2M$ operations). Overall, the *MaxSat* algorithm requires $12M+8A$ operations per pixel.

### 4.4.5.2 SatDesat

The *SatDesat* algorithm performs an additional task of desaturation with $5M + 1A$ operations for each pixel. Hence, including the cost of the *MaxSat*, the total cost of computation becomes $17M + 9A$ per pixel.

### 4.4.5.3   DCT-domain Techniques

As in the DCT domain only the DC coefficient of an $8 \times 8$ block is processed, the computational cost per pixel reduces by a factor of $\frac{1}{64}$. In Table 4.2, the computational costs of different techniques are presented. The expressions take care of the cost of forward and inverse DCT transforms for spatial domain techniques [157].

Table 4.2: Per-pixel computational cost

| Method | Cost |
|---|---|
| MaxSat | $14.75M + 22.5625A$ |
| SatDesat | $19.75M + 23.5625A$ |
| MaxSat-DCT | $\frac{3}{16}M + \frac{1}{8}A$ |
| SatDesat-DCT | $\frac{17}{64}M + \frac{9}{64}A$ |

## 4.5   Color Constancy

From Eq. (4.1), we observe that the color components at a pixel are determined by three main factors, namely, reflection coefficients of the surface point whose image is formed; spectral power density (SPD) of the illuminants or radiations received by the scene from various sources; and the spectral responses of camera sensors. Typically, we consider the imaging of a 3-D scene under varying illumination. In that case, pixel values of images under different scene illumination vary greatly, and the colors rendered by display systems may also show these variations. However, a human observer is capable of perceiving true colors of the objects present in the scene even under this situation. Color constancy of a scene refers to restoration of colors from varying illumination of a scene. Its primary objective is to derive an illumination-independent representation of an image so that it could be suitably rendered with a desired illuminant. The problem has two parts. The first one attempts to estimate the spectral component of the illuminants, and in the other part, it performs color correction for rendering the image with a target illumination.

## 4.5.1  Estimating Spectral Components of a Single Illuminant

Many techniques have been reported [12, 13, 43] for estimating the spectral components of the illuminant of a scene, with the assumption that it has a single illuminant. However, all these techniques solve the color constancy problem in the spatial representation of images in the RGB color space. For adapting them in the compressed domain, we need to perform processing in the YCbCr color space. In a simplified model [43], we assume that all the reflecting bodies have ideal 2-D flat Lambertian surfaces. According to Eq. (4.1), one of the objectives is to estimate $E(\lambda)$ from an input color image, which is expressed in the form of a color vector $(R_E, G_E, B_E)$ in the RGB color space. It is observed [43] that the problem is equivalent to finding a solution to an underconstrained set of equations. Hence, to solve this problem, we have to consider additional assumptions or constraints imposed from the environment. These are briefly discussed below with their necessary adaptation in the YCbCr color space.

1. *Gray World*: In the *gray world assumption* [17, 50], it is assumed that the average reflectance of all surfaces is gray or achromatic. Hence, the average of color components provides the colors of the incident illuminant.

2. *White World*: In this approach [81], an attempt is made to identify a white object in the scene and the color of these white pixels in the image determines the color of the illuminant. To obtain these values, we look for the maximum values of individual color components. This method is quite sensitive to the dynamic ranges of the sensors. However, if the dynamic range of brightness distribution satisfies the linear response of the sensor, this assumption works well in many cases. This assumption is referred to here as the *white world assumption*. In another variation in the YCbCr color space, instead of computing the maximum values in all the three components, the computation is carried out only in the $Y$ component. The corresponding chromatic components of the pixel with maximum $Y$ are chosen for the color of the illuminant. This assumption is referred to as *white world in YCbCr*.

3. *Statistical Techniques*: A more recent trend in solving the color constancy problem is to use statistical estimation techniques with prior knowledge on the distribution of pixels in a color space given known camera sensors and source of illumination. In these techniques, an illuminant (or a set of illuminants) is chosen from a select set of canonical illuminants based on certain criteria. Some of these approaches are discussed below.

   (a) *Color Gamut Mapping*: The color gamut of a scene provides the statistics of color distribution under an illumination. In this technique, an illuminant is mapped to a color gamut covering all possible color renditions under its exposure. There are color gamut mapping approaches both in the 3D [44] and in the 2D [42] color spaces, where an objective is set to maximize the evidence of color maps with known maps of canonical illuminants. In a typical 2-D implementation [109], this statistics is collected in the CbCr space by discretizing it into $32 \times 32$ cells.

(b) *Color-by-Correlation* In this approach [43], the attempt is made to maximize a likelihood of an illuminant given the distribution of pixels in the 2D chromatic space. In [43], it has also been shown that the computational framework is generic enough to accommodate many other algorithms (like the gamut mapping one in the 2D chromatic space). In [43], the chosen chromatic space is formed by *normalized red* $(r = \frac{R}{R+G+B})$ and *normalized green* $(g = \frac{G}{R+G+B})$ components. In the compressed domain [109], the method is adapted in the CbCr space. and statistics is collected in the same way.

(c) *Nearest Neighbor Classification* As the above two techniques require significant amount of storage space for storing the statistics of each canonical illuminant, in [109] a simple *nearest neighbor* (NN) *classification* is used for determining the canonical illuminant. Let $C \in Cb \times Cr$ denote the SPD of an illuminant in the CbCr space, which follows a 2-D Gaussian distribution as given below:

$$p(C) = \frac{1}{2\pi|\Sigma|^{\frac{1}{2}}} e^{-\frac{1}{2}(C-\mu)\Sigma^{-1}(C-\mu)^t}, \tag{4.12}$$

where $\mu(= [\mu_{Cb}\ \mu_{Cr}])$ is the mean of the distribution and $\Sigma(= \begin{bmatrix} \sigma_{Cb}^2 & \sigma_{CbCr} \\ \sigma_{CbCr} & \sigma_{Cr}^2 \end{bmatrix})$ is the covariance matrix . The canonical illuminant is chosen based on the distance of its mean illuminant from the mean chromatic component of the image. The distance function used in [109] is the *Mahalanobis distance function* [91]. Let the mean chromatic components of an image be $C_m$. Then, for an illuminant $L$ with the mean $\mu_L$ and the covariance matrix $\Sigma_L$, the distance between them is given below:

$$d(C_m, \mu_L) = (C_m - \mu_L)\Sigma_L^{-1}(C_m - \mu_L)^T. \tag{4.13}$$

### 4.5.1.1 Computation in the Block DCT Space

For computing in the block DCT space, we simply estimate the color of the illuminant from the DC coefficients of individual blocks, which is the average of $8 \times 8$ sample points in a block. However, with a marginal increase in computation, we include more number of points by taking three more additional points (averages of $4 \times 4$ subblocks) for each block. The same strategy is followed in [113] while collecting statistics for determining color pallets in the block DCT space. This is carried out by performing an inverse 2D DCT directly over the DCT coefficients.

**Table 4.3**: List of algorithms for estimating the color components of an illuminant

| Algorithms | Domain | Short Name |
|---|---|---|
| Gray World | Spatial | GRW |
| Gray World | Block DCT | GRW-DCT |
| White World | Spatial | MXW |
| White World in RGB | Block DCT | MXW-DCT |
| White World in YCbCr | Block DCT | MXW-DCT-Y |
| Color by Correlation | Spatial | COR |
| Color by Correlation | Block DCT | COR-DCT |
| Gamut Mapping | Spatial | GMAP |
| Gamut Mapping | Block DCT | GMAP-DCT |
| Nearest Neighbor | Spatial | NN |
| Nearest Neighbor | Block DCT | NN-DCT |

Let $X_{ij}$, $0 \leq i, j \leq 7$ denote the DCT coefficients of an $8 \times 8$ block. From them the inverse $2 \times 2$ DCTs $y_{ij}$, $0 \leq i, j \leq 1$, are computed as follows.

$$\begin{bmatrix} y_{00} \\ y_{01} \\ y_{10} \\ y_{11} \end{bmatrix} = \frac{1}{2} \begin{bmatrix} 1 & 1 & 1 & 1 \\ 1 & -1 & 1 & -1 \\ 1 & 1 & -1 & -1 \\ 1 & -1 & -1 & 1 \end{bmatrix} \begin{bmatrix} X_{00} \\ X_{01} \\ X_{10} \\ X_{11} \end{bmatrix}. \tag{4.14}$$

As the computation is carried out with a lower resolution of input images, let us discuss its effect in the quality of results. As reported in [109], it is found that the technique provides comparable results in estimating the spectral components with those obtained from the spatial domain techniques dealing with the full resolution of images. To demonstrate the comparative performances of different approaches, let us consider a metric named *angular error* reflecting the quality of estimation of the SPD. Let the target illuminant $T$ be expressed by the spectral component triplet in the RGB color space as $(R_T, G_T, B_T)$, and let the corresponding estimated illuminant be represented by $E = (R_E, G_E, B_E)$. Then, the *angular error* $(\Delta_\theta)$ between these two is defined as

$$\Delta_\theta = cos^{-1}\left(\frac{\vec{T}.\vec{E}}{|\vec{T}||\vec{E}|}\right). \tag{4.15}$$

In the above definitions, '.' denotes the dot product between two vectors, and |.| denotes the magnitude of the vector. The techniques under comparison are listed in Table 4.3.

The SPDs of the illuminants are estimated from the image data set captured by Barnard et al. [14] using different illuminants. These data are available at the website *http://www.cs.sfu.ca/~colour/data*. There are 11 illuminants as listed in Table 4.4. Every image in the data set has also information

Table 4.4: List of illuminants.

| Illuminant | Nature of Source | Short Name |
|---|---|---|
| Philips Ultralume | Fluorescent | ph-ulm |
| Sylvania cool white | Fluorescent | syl-cwf |
| Sylvania warm white | Fluorescent | syl-wwf |
| Sylvania 50MR16Q | Incandescent | syl-50mr16q |
| Same with blue filter | | syl-50mr16q+3202 |
| Lamp at 3500K temperature | Incandescent | solux-3500 |
| Same with blue filter | | solux-3500+3202 |
| Lamp at 4100K temperature | Incandescent | solux-4100 |
| Same with blue filter | | solux-4100+3202 |
| Lamp at 4700K temperature | Incandescent | solux-4700 |
| Same with blue filter | | solux-4700+3202 |

Table 4.5: Performance of different techniques of estimating illuminants

| Technique | $\Delta_\theta$ (in degree) |
|---|---|
| GRW-DCT | 14.77 |
| MXW-DCT | 13.82 |
| MXW-DCT-Y | 14.88 |
| COR-DCT | 6.97 |
| GMAP-DCT | 10.8 |
| NN-DCT | 7.71 |
| GRW | 14.77 |
| MXW | 29.7 |
| COR | 11.45 |
| GMAP | 11.41 |
| NN | 7.1 |

regarding the spectral components of the illuminant in the RGB color space. These are used for collecting statistics for different statistical techniques. In Table, 4.5 the median of the angular errors of the estimates over the complete set of images is shown. We observe that compressed domain techniques performed better than spatial domain techniques in this respect.

### 4.5.1.2   Cost of Computation and Storage

For expressing the computational cost of different techniques, let the number of pixels in an image be denoted by $n$. Moreover, while accounting for the cost of computation, in addition to the number of multiplications and additions, the number of comparisons is also taken into consideration. By extending our previous notation, the computational complexity is expressed as $\alpha M +$

$\beta A + \gamma C$, implying $\alpha$ number of multiplications, $\beta$ number of additions and $\gamma$ number of comparisons. Computational costs for a multiplication and for a division are also considered as equivalent.

For $n$ pixels, the grey-world (GRW) and the white-world (MXW) techniques, respectively, require $3M + 3(n-1)A$ and $3(n-1)C$ number of computations. For statistical techniques, it requires $9nM + 6nA$ operations for converting $n$ pixels in the RGB color space to the YCbCr color space. These techniques also depend on the number of illuminants (say, $n_l$) and the fractional coverage (say, $f$) over the discretized chromaticity space. Let us consider the size of the discretized space as $n_c$, which is mentioned as $32 \times 32$ in algorithmic descriptions (see Section 4.5.1). Hence, the number of computations for processing $n$ pixels becomes $fn_cn_lA + (n_l - 1)C$ for both the techniques COR and GMAP, although the latter performs only integer additions. The NN technique computes the *Mahalanobis distance* with $8M + 4A$ operations for each illuminant, as it uses $2 \times 1$ chromatic mean and symmetric covariance matrices of dimension $2 \times 2$. Hence, the total number of computations for the NN becomes $8n_lM + 4n_lA$ for $n$ pixels.

Every statistical technique needs to store the mean of the illuminant of dimension $3 \times 1$. Moreover, the NN technique needs to store three elements of the covariance matrix per illuminant. It makes its minimum storage requirement as $6n_l$ number of elements. However, both the color-by-correlation (COR) and gamut-mapping (GMAP) techniques require more space for storing the chromaticity map, which has $n_cn_l$ number of elements. Additionally, they also need to store the mean illuminant (3 per illuminant).

In the block DCT space, color constancy techniques handle four $2 \times 2$ IDCT coefficients for each block. The computation of these values, as shown in Eq. (4.14), requires 9 additions with only a marginal overhead in the computation. Due to these operations, the input data size itself gets reduced by $\frac{1}{4}$th of the size of the block, which is 16 times in the context of JPEG compression. Hence, all the nonstatistical-based techniques run faster in the DCT domain. However, the statistical techniques are relatively independent of the input data size. There are savings from the conversion of pixels in the RGB color space to the YCbCr color space. Moreover, in the compressed domain, the additional overhead of reverse and forward transforms to and from the spatial domain is also avoided in DCT-based approaches. A summary of computational and storage complexities of all the techniques is provided in Table 4.6.

## 4.5.2 Color Correction

If we get the estimates of the SPD of the illuminant, this becomes useful in transforming the pixel values to those under a target illuminant. This computation is known as *color correction* and is performed following the Von Kries model [80]. The corresponding method is known as *diagonal color correction*, which is described in the following text. In the block DCT space, the operation is restricted to the DC coefficients only.

**Table 4.6**: Complexities of different algorithms given $n_l$ number of illuminants, $n_c$ as the size of the 2-D chromaticity space, and $n$ number of image pixels

| Algorithms | Computational Complexity | Storage Complexity |
|---|---|---|
| GRW | $3M + 3(n-1)A$ | – |
| GRW-DCT | $3M + 3(\frac{4n+9}{64} - 1)A$ | – |
| MXW | $3(n-1)C$ | – |
| MXW-DCT | $\frac{9}{64}nM + \frac{24n+9}{64}A + 3(\frac{n}{64} - 1)C$ | – |
| MXW-DCT-Y | $(\frac{4n+9}{64} - 1)C$ | – |
| COR | $9nM + (6n + fn_cn_l)A + (n_l - 1)C$ | $n_cn_l + 3n_l$ |
| COR-DCT | $fn_cn_lA + (n_l - 1)C$ | $n_cn_l + 3n_l$ |
| GMAP | $9nM + (6n + fn_cn_l)A + (n_l - 1)C$ | $n_cn_l + 3n_l$ |
| GMAP-DCT | $fn_cn_lA + (n_l - 1)C$ | $n_cn_l + 3n_l$ |
| NN | $(9n + 8n_l)M + (6n + 4n_l)A$ | $6n_l$ |
| NN-DCT | $8n_lM + 4n_lA$ | $6n_l$ |

Let the spectral components for the source illuminant be estimated as $R_s$, $G_s$, and $B_s$ for red, green, and blue spectrum zones of the SPD, respectively. Let the corresponding spectral components for the target illuminant be given by $R_d$, $G_d$, and $B_d$. From a source pixel in the RGB color space with $R$, $G$, and $B$ as its corresponding color components, the transformed or color-corrected components ($R_u$, $G_u$, and $B_u$, respectively) are expressed as follows:

$$k_r = \frac{R_d}{R_s}, \quad k_g = \frac{G_d}{G_s}, \quad k_b = \frac{B_d}{B_s},$$
$$f = \frac{R+G+B}{k_rR+k_gG+k_bB},$$
$$R_u = fk_rR, \quad G_u = fk_gG, \quad B_u = fk_bB. \tag{4.16}$$

### 4.5.2.1 Color Correction in the YCbCr Color Space

The expressions for diagonal corrections (see Eq. (4.16)) in the RGB color space are translated in the YCbCr color space, as it deems to be appropriate for processing in the compressed domain. This is stated in the following theorem.

**Theorem 4.2** *Let $k_r$, $k_g$, and $k_b$ be the parameters for diagonal correction as defined in Eq. (4.16). Given a pixel with color values in the YCbCr color space, the updated color values $Y_u$, $C_{bu}$, and $C_{ru}$ are expressed by the following* *equations:*

$$C'_b = C_b - 128,$$
$$C'_r = C_r - 128,$$
$$f = \frac{3.51Y + 1.63C'_b + 0.78C'_r}{1.17(k_r + k_g + k_b)Y + (2.02k_b - 0.39k_g)C'_b + (1.6k_r - 0.82k_g)C'_r},$$
$$Y_u = f((0.58k_g + 0.12k_b + 0.30k_r)Y + 0.2(k_b - k_g)C'_b + 0.41(k_r - k_g)C'_r),$$
$$C_{bu} = f((0.52k_b - 0.34k_g - 0.18k_r)Y + (0.11k_g + 0.89k_b)C'_b + 0.24(k_g - k_r)C'_r) + 128,$$
$$C_{ru} = f((0.52k_r - 0.43k_g - 0.09k_b)Y + 0.14(k_g - k_b)C'_b + (0.3k_g + 0.7k_r)C'_r) + 128.$$

(4.17)

□

However, there is no significant reduction in the number of multiplications and additions in Eq. (4.17) compared to the diagonal correction method applied in the RGB color space.

#### 4.5.2.2 Color Correction by Chromatic Shift

In [39] a simple color correction method is proposed by translating the chromatic components by the corresponding estimated components of the source illuminant. By adopting this simple strategy, it was demonstrated that good quality color rendition (or transfer) is possible. In this approach, if $Y_d$, $C_{bd}$, and $C_{rd}$ are the color components of a target illuminant in the YCbCr space and the corresponding components in the source illuminant are $Y_s$, $C_{bs}$, and $C_{rs}$, color correction by *chromatic shift* is expressed by the following equations.

$$Y_u = Y,$$
$$C_{bu} = C_b + C_{bd} - C_{bs},$$
$$C_{ru} = C_r + C_{rd} - C_{rs}.$$

(4.18)

Typical examples of color correction following the estimation of the illuminant using the *color-by-correlation* technique in the DCT domain (COR-DCT) are shown in Figure 4.8 (**see color insert**).

---

## 4.6 Color Enhancement

Color enhancement aims at improving the rendering of images and thus increasing their visual perception. There exist a number of techniques [52, 69, 87, 126, 143, 156, 168] in the spatial domain toward this end. In the block DCT space also a few works [1, 84, 105, 141] have been reported in the recent past. Mostly, these techniques attempt to exploit the advantage of spectral factorization in the compressed domain and propose to treat DC and AC coefficients differently for enhancing images. A few representative algorithms are elaborated in subsequent sections.

### 4.6.1    Alpha Rooting

In this technique, smaller values of DCT coefficients (magnitudes less than 1) get emphasized, whereas large values ($> 1$) get reduced. The technique is known as *alpha rooting* (AR) as the coefficients are modified by taking the $(1-\alpha)$th root ($0 < \alpha < 1$) of their normalized values, the normalization factor being the maximum luminance value of the image ($Y_{max}$), which is computed from the DC coefficients in the compressed domain. Thus a DCT coefficient $X(i,j)$ is modified as given by the following equation.

$$\widetilde{X}(i,j) = X(i,j) \left| \frac{X(i,j)}{Y_{max}} \right|^{\alpha-1} . \tag{4.19}$$

### 4.6.2    Multicontrast Enhancement

In this technique (MCE) [141], a contrast measure for spectral bands in the transform domain is defined. Using this measure, the DCT coefficients are modified so that contrasts for individual bands are scaled in the same proportion. In this way, only AC coefficients are modified, without affecting the DC coefficient for each block.

Given the DCT coefficients of a block as $X(i,j), 0 \le i, j \le 7$, a $k$th spectral band is formed by the coefficients lying on the same diagonal such that $i+j = n$. Then the *spectral energy* $E_k$ of the $k$th band is defined as follows:

$$E_k = \frac{1}{N_k} \Sigma_{i+j=k} |X(i,j)|, \tag{4.20}$$

where $N_k$ is the number of coefficients in the $k$th spectral band.

Then the contrast measure ($c_k$) in the $k$th band is defined as the ratio of the spectral energy in that band and the energy of the low-pass-filtered image retaining the frequency components of all the lower bands. This is defined below:

$$c_k = \frac{E_k}{\Sigma_{t<k} E_t}. \tag{4.21}$$

The coefficients of the $k$th band are scaled in such a way that the resulting contrast of the enhanced image becomes $\lambda c_k, 1 \le k \le 14$, where $\lambda$ is a scalar constant. This is carried out by scaling the DCT coefficients of the $k$th band with a factor of $\lambda H_k$, where $H_k$ is defined as

$$H_k = \frac{\Sigma_{t<k} \widetilde{E}_t}{\Sigma_{t<k} E_t}, \tag{4.22}$$

where $\widetilde{E}_t$ and $E_t$ are the spectral energy for the $t$th band in the enhanced and original images, respectively.

### 4.6.3 Multicontrast Enhancement with Dynamic Range Compression

A similar contrast measure is used in this approach also with a minor variation in the definition of the *spectral energy* of a band as given below.

$$E_k = \sqrt{\frac{1}{N_k} \Sigma_{i+j=k} |X(i,j)|^2}. \tag{4.23}$$

Moreover, in this technique, the DC coefficients are transformed with the help of a monotonically increasing mapping function, as it is done in conventional enhancement algorithms such as pixel mapping techniques [52]. The monotonic function (see Figure 4.9 and **its color insert**) used in this algorithm is given below:

$$\eta(x) = \frac{(x^{\frac{1}{\gamma}} + (1 - (1 - x)^{\frac{1}{\gamma}}))}{2}, \quad 0 \le x \le 1. \tag{4.24}$$

The algorithm also varies the parameter ($\gamma$) of dynamic range compression for each block, depending on its spectral content, and subsequently, the parameter is also smoothed by taking the average of its values used in the neighboring blocks. It has also the provision of choosing different *contrast scaling factors* ($\lambda$ in [141]) for low-frequency and high-frequency bands. Further, in [84] a preprocessing step of noise removal and a postprocessing operation of smoothing the DC coefficients are also carried out for further improvement of the result.

### 4.6.4 Color Enhancement by Scaling DCT Coefficients

All the previous algorithms perform enhancement with the luminance component ($Y$) without bringing any change in the corresponding chrominance components $Cb$ and $Cr$. However, in [105], a simple color enhancement technique is proposed by scaling DCT coefficients in all the components. The approach aims at determining a suitable combination of brightness, contrast, and colors of pixels in an image for the betterment of the visualization of images. As it is quite difficult to characterize this combination, it is conjectured that, for natural images an inherent combination of brightness, contrast, and colors should be preserved in the process of enhancement. The *color enhancement by scaling* (CES) technique tries to preserve both the *contrast* and the *color* for enhancing images.

#### 4.6.4.1 Preservation of Contrast

In this approach, a different contrast measure is used from the intuitive notion based on Weber's law. Consider an image of size $N \times N$ with mean $\mu$ and standard deviation $\sigma$. According to Weber's law, the contrast ($\zeta$) of an image is given by $\zeta = \frac{\Delta L}{L}$, where $\Delta L$ is the difference in luminance between a stimulus and its surround, and $L$ is the luminance of the surround [139]. As the mean provides a measure for the surrounding luminance and the standard deviation is strongly correlated with $\Delta L$, the contrast $\zeta$ of an image is redefined as follows:

$$\zeta = \frac{\sigma}{\mu}. \tag{4.25}$$

The following theorem [105] states how the contrast of an image is related to the scaling of its DCT coefficients.

**Theorem 4.3** *Let $\kappa_{dc}$ be the scale factor for the normalized DC coefficient, and $\kappa_{ac}$ be the scale factor for the normalized AC coefficients of an image $Y$ of size $N \times N$, such that the DCT coefficients in the processed image $\widetilde{Y}$ are given by*

$$\widetilde{Y}(i,j) = \begin{cases} \kappa_{dc} Y(i,j), & \text{for } i = j = 0, \\ \kappa_{ac} Y(i,j), & \text{otherwise.} \end{cases} \tag{4.26}$$

*The contrast of the processed image then becomes $\frac{\kappa_{ac}}{\kappa_{dc}}$ times of the contrast of the original image.*

□

Hence, for preserving the local contrast of each block in the block DCT space, $\kappa_{dc}$ and $\kappa_{ac}$ should be the same. Let us denote the corresponding scaling factor as $\kappa$.

### 4.6.4.2 Preservation of Color

Though the above operations in the luminance component of an image preserve the contrast, they do not preserve the colors or color vectors of the pixels. Hence, we need to adjust also chromatic components. The following theorem [105] states how colors could be preserved under the uniform scaling operation.

**Theorem 4.4** *Let $U = \{U(k,l)|0 \le k,l \le (N-1)\}$ and $V = \{V(k,l)|0 \le k,l \le (N-1)\}$ be the DCT coefficients of the Cb and Cr components, respectively. If the luminance ($Y$) component of an image is uniformly scaled by a factor $\kappa$, the colors of the processed image with $\widetilde{Y}$, $\widetilde{U}$ and $\widetilde{V}$ are preserved by the following operations:*

$$\widetilde{U}(i,j) = \begin{cases} N(\kappa(\frac{U(i,j)}{N} - 128) + 128), & i = j = 0, \\ \kappa U(i,j), & \text{otherwise,} \end{cases} \tag{4.27}$$

$$\widetilde{V}(i,j) = \begin{cases} N(\kappa(\frac{V(i,j)}{N} - 128) + 128), & i = j = 0, \\ \kappa V(i,j), & \text{otherwise.} \end{cases} \tag{4.28}$$

□

### 4.6.4.3 The Algorithm

The CES algorithm has been developed based on the above two theorems. Its different stages of computation are described below.

1.  **Adjustment of local background illumination:** For adjusting the background illumination, the DC coefficients are scaled using a global monotonic function. In [105], the following function is reported to be quite effective for performing this task. The function (see Figure 4.10 and **its color insert**), known as the *twicing function*, is also used in other works [101] for the purpose of enhancement.

$$\tau(x) = x(2 - x), \quad 0 \le x \le 1. \tag{4.29}$$

    Once a scaling factor is determined from the DC coefficient of the block, the same is used for scaling other DCT coefficients in subsequent operations.

2.  **Preservation of local contrast:** According to Theorem 4.3, to preserve the contrast of the block, AC coefficients are also scaled with the same factor of the DC coefficient. However, it may cause overflow of values from the displayable range of the device. Hence, to restrict this overflow, the scale factor is clipped by a value as determined by the following theorem [105].

    **Theorem 4.5** *Assuming that the values in a block lie within* $\mu \pm \lambda \sigma$*, the scaled values will not exceed the maximum displayable brightness value* $B_{max}$*, if* $1 \le \kappa \le \frac{B_{max}}{\mu + \lambda.\sigma}$*.*

    □

3.  **Preservation of colors:** In the last step, colors are preserved by scaling the DCT coefficients of the $Cb$ and $Cr$ components according to Theorem 4.4.

It is observed that, in many cases, blocking artifacts occur near edges or near sharp changes of brightness and colors of pixels. Hence, in those cases, it is preferred to compute over a smaller block size. This is efficiently performed by a sequence of block decomposition, processing in smaller blocks, and finally composing the processed blocks into its original size. As discussed in Chapter 2 (see Section 2.2.4.5), the block decomposition and composition operations are performed efficiently in the DCT domain. To detect a block requiring these additional operations, the *standard deviation* $\sigma$ of that block is used as a measure. If its value is greater than a threshold (say, $\sigma_{th}$), the $8 \times 8$ block is decomposed into *four* subblocks, and the scaling operations are applied in those blocks independently.

**Table 4.7**: Average performance measures obtained by different color enhancement techniques

| Measures | AR | MCE | MCEDRC | CES |
|---|---|---|---|---|
| JPQM | 9.9 | 8.32 | 9.71 | 9.03 |
| CEF | 1.00 | 0.93 | 1.00 | 1.51 |

**Table 4.8**: Performance measures after iterative application of the CES algorithm on the image 'image20'

| Measures | Number of iterations | | | |
|---|---|---|---|---|
| | 2 | 3 | 4 | 5 |
| CEF | 1.96 | 2.21 | 2.24 | 2.23 |
| JPQM | 8.05 | 7.32 | 6.69 | 6.37 |

### 4.6.5    Examples of Color Enhancement

Let us consider a few typical examples of color enhancement using the techniques as discussed earlier. In Figure 4.11 (**see color insert**), two such images are shown[2]. In the implementation of the AR technique, the value of $\alpha$ has been kept as 0.98. In the MCE, the value of $\lambda$ is kept at 1.95. In *multicontrast enhancement with dynamic range compression* (MCEDRC), the parameter $\gamma$ of the function $\eta(x)$ is empirically chosen as 1.95. The average JPQM and CEF values obtained on this set of images are presented in Table 4.7.

From Table 4.7, the CES method is found to provide excellent color rendition in the enhanced images. However, as indicated from the JPQM measures, its performance in suppressing blocking artifacts is marginally poorer than that of other schemes. In Figure 4.12 (**see color insert**), enhancement results obtained by these techniques are presented.

#### 4.6.5.1    Iterative Enhancement

We also observe the effect of iterative enhancement of an image where the enhanced image from the previous iteration acts as the input to the next stage of enhancement. It is found that the CEF improves with a few number of iterations. However, the resulting images heavily suffer from blocking artifacts. In Table 4.8, CEF and JPQM values are shown on the iterative application of the CES algorithm over the image 'image20.' Results after 2 and 3 iterations are also shown in Figure 4.13 (**see color insert**).

---

[2]Obtained from the website *http://dragon.larc.nasa.govt/retinex/pao/news*

## 4.7 Summary

Color processing in the compressed domain has the advantage of getting the input image in the YCbCr color space where its chromatic components are separated from the luminance component. This property is exploited in various techniques in the compressed domain by restricting the computation to the relevant components only. Moreover, the spectral factorization of the data in the DCT space provides other opportunities for designing efficient and effective algorithms by treating the DC and AC coefficients of each block differently. A few representative problems of color processing, namely, the saturation and desaturation of colors, color constancy, and color enhancement are discussed to highlight key approaches in the block DCT space.

Original Image

(a)

Saturation in Nomralized CbCr

(b)

**Figure 4.4**: Maximum saturation in nCbCr: (a) image22 and (b) maximally saturated image.

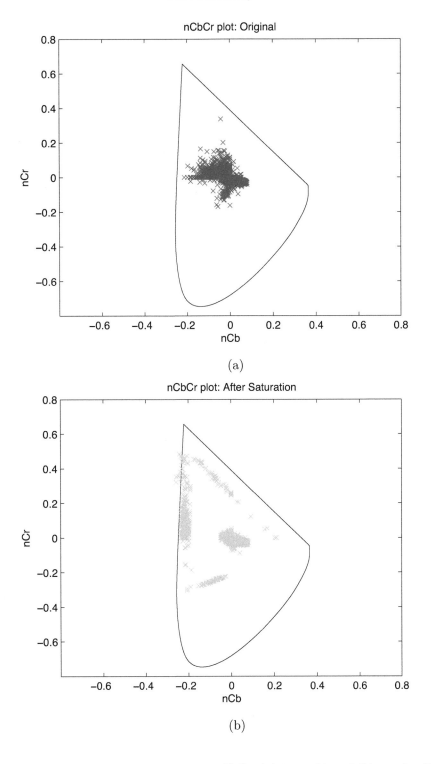

**Figure 4.5**: Chromaticity points in nCbCr: (a) image22 and (b) maximally saturated image.

Saturation (nCbCr)–Desaturation

(a)

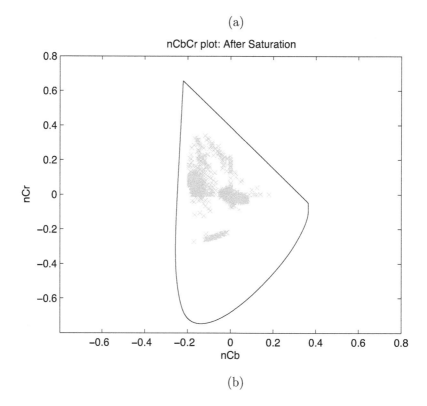

(b)

**Figure 4.6**: (a) image22: Desaturation from the maximally saturated image and (b) corresponding chromaticity plot in the nCbCr space.

Saturation in Nomralized CbCr

(a)

Saturation (nCbCr)–Desaturation

(b)

Figure 4.7: image22: (a) Maxsat in DCT and (b) SatDesat in DCT.

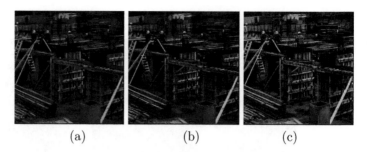

(a)                    (b)                    (c)

**Figure 4.8**: Example of color correction using COR-DCT: (a) original image, (b) diagonal correction, and (c) chromatic shift.

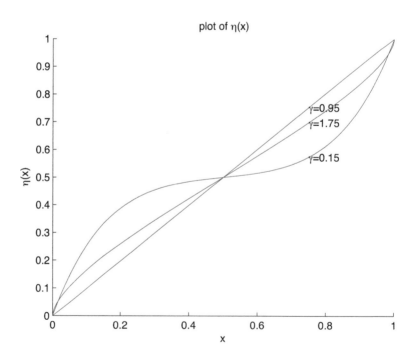

Figure 4.9: Plot of $\eta(x)$.

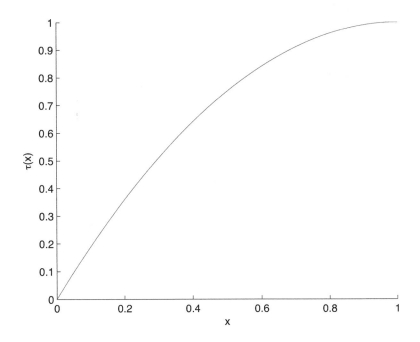

Figure 4.10: Plot of $\tau(x)$.

(a)            (b)

Figure 4.11: Original images: (a) image7 and (b) image20.

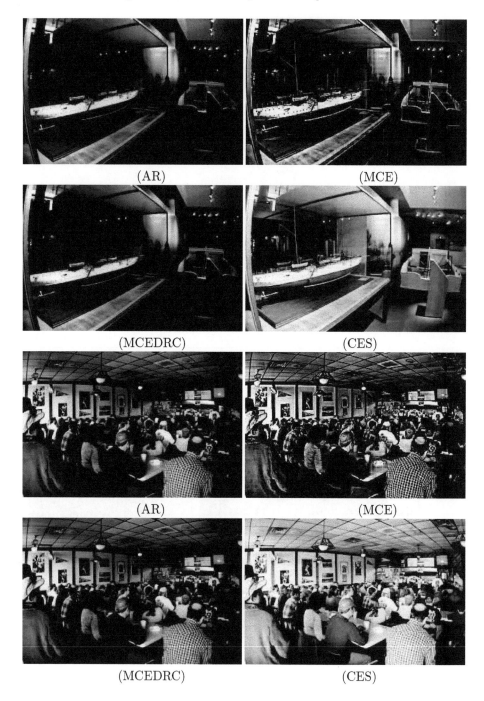

(AR)            (MCE)

(MCEDRC)            (CES)

(AR)            (MCE)

(MCEDRC)            (CES)

Figure 4.12: Color enhancement of the images.

(a)                                      (b)

**Figure 4.13**: Color enhancement of 'image20' obtained with more than one iteration: (a) 2 and (b) 3.

(a)                                          (b)

**Figure 1.28:** Color image enhancement in the block DCT space: (a) Original, and (b) Enhanced image.

Original Image

(a)

Saturation in Nomralized CbCr

(b)

**Figure 4.4: Maximum saturation in nCbCr: (a) image22, and (b) maximally saturated image.**

## Saturation (nCbCr)–Desaturation

(a)

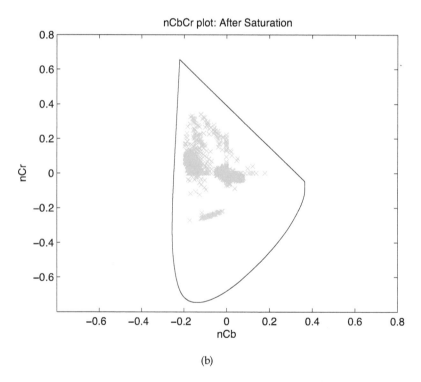

(b)

Figure 4.6: (a) image22: Desaturation from the maximally saturated image, and (b) corresponding chromaticity plot in the nCbCr space.

Saturation in Nomralized CbCr

(a)

Saturation (nCbCr)–Desaturation

(b)

Figure 4.7: image22: (a) Maxsat in DCT, and (b) SatDesat in DCT.

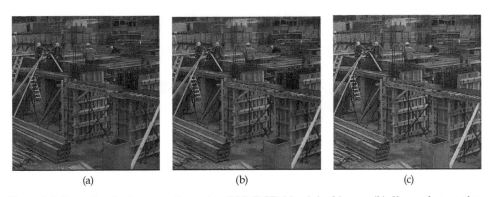

(a)                              (b)                              (c)

Figure 4.8: Example of color correction using COR-DCT: (a) original image, (b) diagonal correction, and (c) chromatic shift.

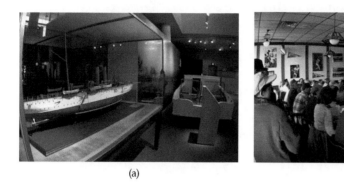

(a)                                        (b)

Figure 4.11: Original images: (a) image7, and (b) image20.

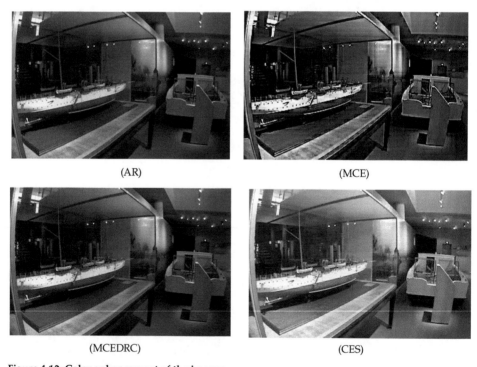

(AR)                                       (MCE)

(MCEDRC)                                    (CES)

Figure 4.12: Color enhancement of the images.

(*continued*)

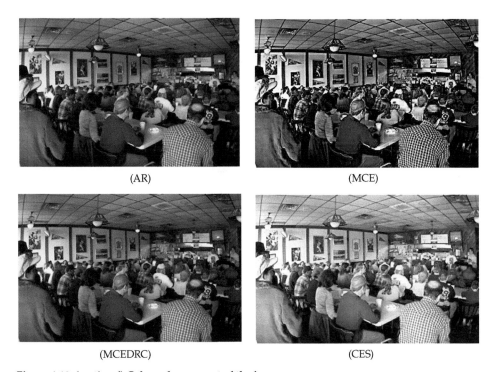

(AR)  (MCE)

(MCEDRC)  (CES)

Figure 4.12: (*continued*) Color enhancement of the images.

(a)  (b)

Figure 4.13: Color enhancement of `image20' obtained with more than one iteration: (a) 2, and (b) 3.

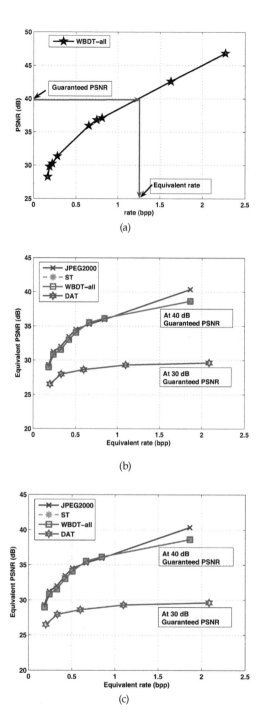

Figure 6.2: JPEG2000 to JPEG transcoding for Lena image using the WBDT technique: (a) rate distortion for transcoded image with respect to the JPEG2000 decoded image (for 0.55 bpp encoded Lena image), (b)equivalent rate of the transcoder versus JPEG2000 compression rate, and (c)equivalent PSNR versus equivalent rate. (From [147] with permission from the author.)

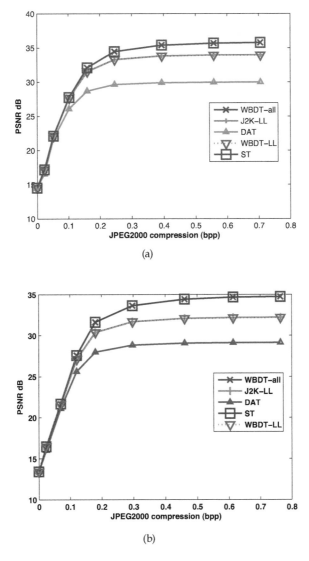

(a)

(b)

Figure 6.3: Maximum PSNR values versus compression ratio:(a) Lena, (b) Peppers, (c) Girl, and (d) Baboon. (With permission from [147].)

(*continued*)

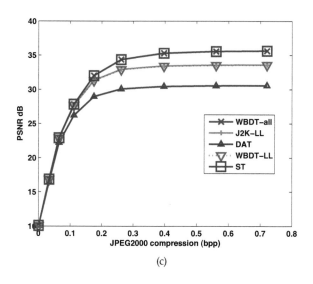

(c)

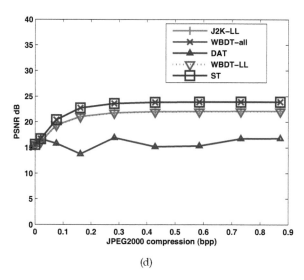

(d)

Figure 6.3: (*continued*) Maximum PSNR values versus compression ratio:(a) Lena, (b) Peppers, (c) Girl, and (d) Baboon. (With permission from [147].)

# Chapter 5

## Image Resizing

Image resizing is often required for accommodating images into a suitable format as demanded by different applications involving display, transmission, and storage of images. Display devices over a wide range of resolution are available nowadays. Camera resolution also varies widely, and there is a need for displaying images in a desired spatial resolution in agreement with display resolutions. Hence, resizing is a task that needs to be performed in such a scenario. This is true not only for images, but videos also need to be resized for the same purpose. However, in this chapter we restrict ourselves to the problem of image resizing. The problem of video resizing is more complex and is closely related to the more general problem of transcoding of images and videos. Hence, this is discussed in Chapter 6.

The problem of image resizing deals with conversion of an image of size $M_1 \times N_1$ to that of a different size, $M_2 \times N_2$. In particular, when $M_2 = \frac{M_1}{2}$ and $N_2 = \frac{N_1}{2}$, the operation is known as *image halving*. On the other hand, for $M_2 = 2M_1$ and $N_2 = 2N_1$, the process is referred to as *image doubling*. If both $M_2$ and $N_2$ are less than $M_1$ and $N_1$, respectively, the process is called *downsampling*. Again if they are greater than the other two, it is known as *upsampling*. There are various interpolation techniques for performing this task in the spatial domain. In this chapter, we review some of the key approaches

[36, 60, 62, 94, 99, 108, 103, 106, 111, 123, 124, 132, 134, 136] to solving this problem in the block DCT space by exploiting different properties of DCT as discussed in Chapter 2. In the next section, we first consider the problem of image halving and image doubling. Subsequently, techniques to carry out more general resizing tasks are discussed.

## 5.1   Image Halving and Image Doubling in the Compressed Domain

In the early stage of development of various techniques in the transform domain, problems of image halving and image doubling drew considerable attention from various researchers. Later, some of these techniques were extended to the development of resizing algorithms with arbitrary factors. To understand the development of these concepts, we first review various approaches to image halving and image doubling in the following subsections.

### 5.1.1   Using Linear, Distributive and Unitary Transform Properties

Several approaches use the linear, distributive, and unitary transform properties of the DCT for resizing images in this domain. For example, in [123] a simple algorithm of image halving is reported that exploits those properties. In this approach, adjacent four $8 \times 8$ blocks are converted into one block, which contains averages of $2 \times 2$ sub-blocks of pixels from each of them.

Let $x_{ij}, 0 \leq i, j \leq 1$, denote these adjacent four blocks in the spatial domain (as shown in Figure 5.1). The downsampled block $x_d$ is generated from these blocks according to the following Eq. (5.1).

$$x_d = \Sigma_{j=0}^{1} \Sigma_{i=0}^{1} p_i x_{ij} p_j^T, \qquad (5.1)$$

where,

$$p_0 = \begin{bmatrix} D_{4 \times 8} \\ 0_{4 \times 8} \end{bmatrix}, \quad p_1 = \begin{bmatrix} 0_{4 \times 8} \\ D_{4 \times 8} \end{bmatrix}. \qquad (5.2)$$

In the above equations, $0_{4 \times 8}$ is a $4 \times 8$ null (or zero) matrix, and $D_{4 \times 8}$ is defined as

$$D_{4 \times 8} = \begin{bmatrix} 0.5 & 0.5 & 0 & 0 & 0 & 0 & 0 & 0 \\ 0 & 0 & 0.5 & 0.5 & 0 & 0 & 0 & 0 \\ 0 & 0 & 0 & 0 & 0.5 & 0.5 & 0 & 0 \\ 0 & 0 & 0 & 0 & 0 & 0 & 0.5 & 0.5 \end{bmatrix}. \qquad (5.3)$$

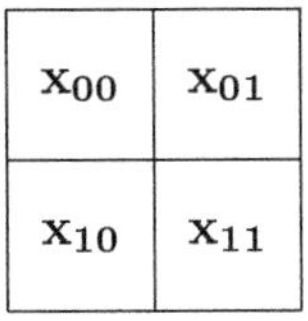

| $x_{00}$ | $x_{01}$ |
|----------|----------|
| $x_{10}$ | $x_{11}$ |

Figure 5.1: Four adjacent spatial domain blocks.

In the transform domain, Eq. (5.1) is given by

$$DCT(\mathbf{x_d}) = \Sigma_{j=0}^1 \Sigma_{i=0}^1 DCT(p_i) DCT(\mathbf{x_{ij}}) DCT(p_j^T). \qquad (5.4)$$

Eq. (5.4) provides the equivalent computation of bilinear decimation technique in the transform domain. A typical result from the transform domain operations is shown in Figure 5.2. Even though the matrices $p_1$ and $p_2$ are sparse in the spatial domain, their DCTs (denoted by $P_1$ and $P_2$, respectively) are not so sparse, as shown below.

$$P1 = \begin{bmatrix} 0.5 & 0 & 0 & 0 & 0 & 0 & 0 & 0 \\ 0.453 & 0.204 & -0.034 & 0.010 & 0 & -0.006 & 0.014 & -0.041 \\ 0 & 0.49 & 0 & 0 & 0 & 0 & 0 & -0.098 \\ -0.159 & 0.388 & 0.237 & -0.041 & 0 & 0.027 & -0.098 & -0.077 \\ 0 & 0 & 0.462 & 0 & 0 & 0 & -0.191 & 0 \\ 0.106 & -0.173 & 0.355 & 0.204 & 0 & -0.136 & -0.147 & 0.034 \\ 0 & 0 & 0 & 0.416 & 0 & -0.278 & 0 & 0 \\ -0.090 & 0.136 & -0.173 & 0.360 & 0 & -0.240 & 0.072 & -0.027 \end{bmatrix}, \qquad (5.5)$$

$$P2 = \begin{bmatrix} 0.500 & 0 & 0 & 0 & 0 & 0 & 0 & 0 \\ -0.453 & 0.204 & 0.034 & 0.010 & 0 & -0.006 & -0.014 & -0.041 \\ 0 & -0.49 & 0 & 0 & 0 & 0 & 0 & 0.098 \\ 0.159 & 0.388 & -0.237 & -0.041 & 0 & 0.027 & 0.098 & -0.077 \\ 0 & 0 & 0.462 & 0 & 0 & 0 & -0.191 & 0 \\ -0.106 & -0.173 & -0.355 & 0.204 & 0 & -0.136 & 0.147 & 0.034 \\ 0 & 0 & 0 & -0.416 & 0 & 0.278 & 0 & 0 \\ 0.090 & 0.136 & 0.173 & 0.360 & 0 & -0.240 & -0.072 & -0.027 \end{bmatrix}. \qquad (5.6)$$

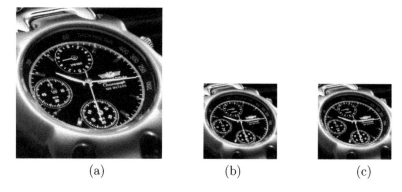

(a)                          (b)                          (c)

**Figure 5.2**: Image halving using linear distributive and unitary properties of DCT: (a) original image, (b) bilinear decimation in spatial domain, and (c) downsampled image in the transform domain (58.24 dB with respect to the image in (b) with JPQM as 8.52).

Hence, if a fast DCT (and IDCT) algorithm [157] is employed in spatial domain processing, it requires less computation compared to the above approach.

## 5.1.2   Using Convolution-Multiplication Properties

In this approach similar to the resizing operations in spatial domain, images are subjected to low-pass filtering before downsampling or after upsampling. In Chapter 3, we have already discussed how filtering is performed directly in the DCT domain by using the convolution multiplication property. For example, in [94], during the downsampling operation, low-pass filtering is applied using the third relationship of Eq. (2.94) of Chapter 2. The relevant relationship in 1-D is restated below.

$$C_{1e}\left(x \circledast y(n-1)\right) = \begin{array}{c} \sqrt{2N} C_{2e}\left(x(l)\right) C_{2e}\left(y(m)\right), \\ 0 \le l, m \le N-1 \ , \ -1 \le n < N. \end{array} \tag{5.7}$$

Using the above, we compute the filtered output in the compressed domain. Next, by applying the downsampling property of DCT coefficients (refer to Theorem 2.11 in Section 2.2.4.2 of Chapter 2), the downsampled coefficients in the type-II DCT space are computed. For the doubling operation, the DCT coefficients are computed using Theorem 2.12 (refer to Section 2.2.4.2), which is followed by a low-pass filtering in the DCT domain. In [132], these techniques are further refined by considering the spatial relationship of adjacent blocks. It should be mentioned here that truncation or zero-padding of DCT coefficients itself is a kind of low-pass filtering operation, which also takes care of the adjustment of coefficients through the upsampling or downsampling processes as described in Theorems 2.11 and 2.12 of Chapter 2.

In [62], instead of a convolution-multiplication property of DCTs, a multiplication-convolution property is used. In this case, downsampling and upsampling operations are shown as the sum of multiplication operations of the sample values with given windows in the spatial domain. The multiplication operations are efficiently computed in the transform domain using symmetric convolution in the DCT domain (refer to the fourth relationship of Eq. (2.94) in Chapter 2). For the sake of brevity, the concept is discussed for processing DCT blocks in 1-D in the following subsections.

### 5.1.2.1 Two-fold Downsampling of 8-point DCT Blocks in 1-D

Let us consider two adjacent blocks of 8 sample points in the spatial domain and denote the sequences as $x_1(n), 0 \le n \le 7$ and $x_2(n), 0 \le n < 7$. Consider the half-symmetric extension of these two sequences such that for $x_1(n)$ the point of symmetry lies in its beginning and the same at the end for $x_2(n)$. The extended sequences are represented in the following forms:

$$\widetilde{x_1}(n) = \begin{cases} x_1(n) & 0 \le n \le 7 \\ x_1(15 - n) & 8 \le n \le 15 \end{cases} \tag{5.8}$$

$$\widetilde{x_2}(n) = \begin{cases} x_2(7 - n) & 0 \le n \le 7 \\ x_2(n - 8) & 8 \le n \le 15 \end{cases} \tag{5.9}$$

This makes the length of the extended sequences 16. To form a concatenated sequence of $x_1(n)$ followed by $x_2(n)$, we multiply them them window functions $w_1(n)$ and $w_2(n)$, respectively, as given in the following expression:

$$x(n) = w_1(n)x_1(n) + w_2(n)x_2(n) \tag{5.10}$$

where

$$w_1(n) = \begin{cases} 1 & 0 \le n \le 7 \\ 0 & 8 \le n \le 15 \end{cases} \tag{5.11}$$

and,

$$w_2(n) = \begin{cases} 0 & 0 \le n \le 7 \\ 1 & 8 \le n \le 15 \end{cases} \tag{5.12}$$

In the DCT domain, equivalent operations on $X_1(k)$ and $X_2(k)$ (DCTs of $x_1(n)$ and $x_2(n)$), respectively, are shown in the following expressions. First, DCTs of extended sequences are obtained as follows:

$$\widetilde{X_1}(k) = \begin{cases} 2X_1(\frac{k}{2}) & \text{for k even.} \\ 0 & \text{for k odd.} \end{cases} \tag{5.13}$$

$$\widetilde{X_2}(k) = \begin{cases} (-1)^{\frac{k}{2}} 2X_1(\frac{k}{2}) & \text{for k even.} \\ 0 & \text{for k odd.} \end{cases} \tag{5.14}$$

In the above equations, both $\widetilde{X_1}(k)$ and $\widetilde{X_2}(k)$ are 16-point DCTs of $x_1(n)$ and $x_2(n)$. Hence, the equivalent concatenation operation as described in Eq.

(5.10) is performed using the convolution multiplication theorem in the following form:

$$X(k) = \frac{1}{4\sqrt{2}}(\widetilde{X_1}(k)\circledS W_1(k) + \widetilde{X_2}(k)\circledS W_2(k)), \tag{5.15}$$

where $W_1(k)$ and $W_2(k)$ are DCTs of $w_1(n)$ and $w_2(n)$. The symbol $\circledS$ denotes the skew circular convolution. Finally, for downsampling 16-point $X(k)$ to 8-point coefficients, we perform the truncation operation with scaling as discussed in Chapter 2. Hence, representing DCT coefficients by column vectors $X_d$ and $X$, respectively, the downsampled DCT coefficients are given by

$$X_d = \begin{bmatrix} I_8 & 0_8 \end{bmatrix} X. \tag{5.16}$$

In [62], how the above computation could be performed efficiently by identifying several redundancies has been discussed. From Eqs. (5.15) and (5.16) equivalent computation is expressed as follows [62].

$$X_d = \frac{1}{4\sqrt{2}}(W_1^d X_1 + S W_1^d S X_2), \tag{5.17}$$

where $W_1^d$ is defined from $W_1$, the DCT of $w_1(n)$, as follows:

$$W_1^d = \begin{bmatrix}
\sqrt{2}W_1(1) & 0 & 0 & 0 & \cdots \\
W_1(2) & W_1(2)+W_1(4) & W_1(4)+W_1(6) & W_1(6)+W_1(8) & \cdots \\
0 & \sqrt{2}W_1(1) & 0 & 0 & \cdots \\
W_1(4) & W_1(2)+W_1(6) & W_1(2)+W_1(8) & W_1(4)+W_1(10) & \cdots \\
0 & 0 & \sqrt{2}W_1(1) & 0 & \cdots \\
W_1(6) & W_1(4)+W_1(8) & W_1(2)+W_1(10) & W_1(2)+W_1(12) & \cdots \\
0 & 0 & 0 & \sqrt{2}W_1(1) & \cdots \\
W_1(8) & W_1(6)+W_1(10) & W_1(4)+W_1(12) & W_1(2)+W_1(14) & \cdots
\end{bmatrix}$$

$$\begin{bmatrix}
\cdots & 0 & 0 & 0 & 0 \\
\cdots & W_1(8)+W_1(10) & W_1(10)+W_1(12) & W_1(12)+W_1(14) & W_1(14)+W_1(16) \\
\cdots & 0 & 0 & 0 & 0 \\
\cdots & W_1(6)+W_1(12) & W_1(8)+W_1(14) & W_1(10)+W_1(16) & W_1(12)-W_1(16) \\
\cdots & 0 & 0 & 0 & 0 \\
\cdots & W_1(4)+W_1(14) & W_1(6)+W_1(16) & W_1(8)-W_1(16) & W_1(10)-W_1(14) \\
\cdots & 0 & 0 & 0 & 0 \\
\cdots & W_1(2)+W_1(16) & W_1(4)-W_1(16) & W_1(6)-W_1(14) & W_1(8)-W_1(12)
\end{bmatrix}. \tag{5.18}$$

In the above equation, due to the constraint of space, the first four columns of each row are shown in the top rows of the matrix and its bottom eight rows contain the remaining four columns corresponding to each of them. Further, $S$ in Eq. (5.17) is defined below:

$$S = \begin{bmatrix}
1 & 0 & 0 & 0 & 0 & 0 & 0 & 0 \\
0 & -1 & 0 & 0 & 0 & 0 & 0 & 0 \\
0 & 0 & 1 & 0 & 0 & 0 & 0 & 0 \\
0 & 0 & 0 & -1 & 0 & 0 & 0 & 0 \\
0 & 0 & 0 & 0 & 1 & 0 & 0 & 0 \\
0 & 0 & 0 & 0 & 0 & -1 & 0 & 0 \\
0 & 0 & 0 & 0 & 0 & 0 & 1 & 0 \\
0 & 0 & 0 & 0 & 0 & 0 & 0 & -1
\end{bmatrix}. \tag{5.19}$$

### 5.1.2.2 Twofold Upsampling of 8-point DCT Blocks in 1-D

While upsampling an 8-point DCT block $X(k), 0 \leq k \leq 7$, first it is converted into a 16-point block by zero-padding and appropriate scaling to $\widetilde{X}(k)$ as follows:

$$\widetilde{X}(k) = \begin{cases} \sqrt{2}X(k) & 0 \leq k \leq 7 \\ 0 & 8 \leq k \leq 15 \end{cases} \tag{5.20}$$

Let the upsampled block in the spatial domain be $x(n)$. We obtain two 8-point blocks by using the same window functions in the following way:

$$\begin{aligned} y_1(n) &= x(n)w_1(n), & 0 \leq n \leq 7, \\ y_2(n) &= x(n+8)w_2(n+8), & 0 \leq n \leq 7. \end{aligned} \tag{5.21}$$

Equivalent operations in DCT space to obtain the DCT of $y_1(n)$ and $y_2(n)$ are shown in the following expressions:

$$\begin{aligned} Y_1(k) &= \tfrac{1}{4}(\widetilde{X}(2k)\textcircled{s}W_1(2k)), & 0 \leq k \leq 7, \\ Y_2(k) &= (-1)^k\tfrac{1}{4}(\widetilde{X}(2k)\textcircled{s}W_2(2k)), & 0 \leq k \leq 7. \end{aligned} \tag{5.22}$$

The above equations use the relationship of DCT coefficients between the $n$-point DCT coefficients with their upsampled versions (of $2n$-point DCT) when the sequence is extended in the spatial domain with trailing zeroes or leading zeroes. The above computation is expressed in matrix notation as follows [62]:

$$\begin{aligned} Y_1 &= \tfrac{1}{2\sqrt{2}}(W_1^u X), \\ Y_2 &= \tfrac{1}{2\sqrt{2}}(SW_1^u SX). \end{aligned} \tag{5.23}$$

where $W_1^u$ is defined from $W_1$ as follows:

$$W_1^u = \begin{bmatrix} \sqrt{2}W_1(1) & 2W_1(2) & 0 & 2W_1(4) & \cdots \\ 0 & W_1(2)+W_1(4) & \sqrt{2}W_1(1) & W_1(2)+W_1(6) & \cdots \\ 0 & W_1(4)+W_1(6) & 0 & W_1(2)+W_1(8) & \cdots \\ 0 & W_1(6)+W_1(8) & 0 & W_1(4)+W_1(10) & \cdots \\ 0 & W_1(8)+W_1(10) & 0 & W_1(6)+W_1(12) & \cdots \\ 0 & W_1(10)+W_1(12) & 0 & W_1(8)+W_1(14) & \cdots \\ 0 & W_1(12)+W_1(14) & 0 & W_1(10)+W_1(16) & \cdots \\ 0 & W_1(14)+W_1(16) & 0 & W_1(12)-W_1(16) & \cdots \\ \cdots & 0 & 2W_1(6) & 0 & 2W_1(8) \\ \cdots & 0 & W_1(4)+W_1(8) & 0 & W_1(6)+W_1(10) \\ \cdots & \sqrt{2}W_1(1) & W_1(2)+W_1(10) & 0 & W_1(4)+W_1(12) \\ \cdots & 0 & W_1(2)+W_1(12) & \sqrt{2}W_1(1) & W_1(2)+W_1(14) \\ \cdots & 0 & W_1(4)+W_1(14) & 0 & W_1(2)+W_1(16) \\ \cdots & 0 & W_1(6)+W_1(16) & 0 & W_1(4)-W_1(16) \\ \cdots & 0 & W_1(8)-W_1(16) & 0 & W_1(6)-W_1(14) \\ \cdots & 0 & W_1(10)-W_1(14) & 0 & W_1(8)-W_1(12) \end{bmatrix} \tag{5.24}$$

$W_1^u$ also follows the same notation of $W_1^d$ regarding representation of rows.

### 5.1.2.3 Example in 2-D

The above concept is extended in 2-D by exploiting the separability property of DCT. Let $X_{i,j}, 1 \leq i, j \leq 2$ be four adjacent $8 \times 8$ DCT blocks. Eq. (5.17) is extended to 2-D in the following form:

$$X_d = \frac{1}{32}(P_d X_{11} P_d^T + P_d X_{12} Q_d^T + Q_d X_{21} P_d^T + Q_d X_{22} Q_d^T), \tag{5.25}$$

(a)                                                    (b)

**Figure 5.3**: Image halving and doubling using convolution-multiplication property [62]: (a) downsampled image(36 dB with respect to the image in Figure 5.2(b) with JPQM as 9.31) and (b) upsampled image with JPQM as 6.29.

where $P_d = W_1^d$ and $Q_d = SW_1^d S$. Similarly, Eq. (5.23) is extended to 2-D as given below:

$$
\begin{array}{rcl}
Y_{11} & = & \frac{1}{8}(P_u X P_u^T), \\
Y_{12} & = & \frac{1}{8}(P_u X Q_u^T), \\
Y_{21} & = & \frac{1}{8}(Q_u X P_u^T), \\
Y_{22} & = & \frac{1}{8}(Q_u X Q_u^T),
\end{array}
\tag{5.26}
$$

where $P_u = W_1^u$ and $Q_u = SW_1^u S$. Typical examples of image halving and image doubling are shown in Figure 5.3. We refer to these techniques in this text as *image halving using convolution multiplication property* (IHCMP) and *image doubling using convolution multiplication property* (IDCMP), respectively.

### 5.1.3   Using Subband DCT Approximation with Block Composition and Decomposition

In some schemes, subband relationships of DCT coefficients (refer to Eqs. (2.165) and (2.166) of Chapter 2) are more effectively used to downsample and upsample input DCT blocks. As higher-order DCT coefficients are ignored here, the operation becomes equivalent to a low-pass filtering operation in the transform domain. The performance of such approximations is found to be better than the conventional low-pass filtering operations [36, 108]. The approximated coefficients are retargeted in DCT blocks of desired sizes (e.g., $8 \times 8$ for JPEG compressed images) with block composition and decomposition. Various intuitive approaches [36, 106, 125] are put forward in this regard. A few representative schemes are highlighted below.

### 5.1.3.1 Image Halving

Consider four adjacent $8 \times 8$ DCT blocks $X_{ij}, 0 \leq i, j \leq 1$. We obtain $4 \times 4$ approximated DCT blocks using the subband relationship from each of them. For simplicity we use here only truncated approximations from Eq. (2.166) in Chapter 2. Let the approximated $4 \times 4$ blocks be represented as $X_{ij}^d$, which is given by the following expression:

$$X_{ij}^d = \frac{1}{2} [X_{ij}(k, l)]_{0 \leq k, l \leq 3}. \qquad (5.27)$$

Next these four decimated blocks are merged to an $8 \times 8$ DCT block (Figure 5.4) with the help of the block composition matrix $A_{(2,4)}$ (refer to Section 2.2.4.5 of Chapter 2).

$$Y = A_{(2,4)} \begin{bmatrix} X_{00}^d & X_{01}^d \\ X_{10}^d & X_{11}^d \end{bmatrix} A_{(2,4)}^T. \qquad (5.28)$$

As the above technique performs the operation through subband approximation followed by block composition, we refer to this approach as *Image Halving through Approximation followed by Composition* (IHAC) [103]. The above computation could be performed more efficiently by exploiting the sparse property of $A_{(2,4)}$ and also the relationship among the elements of its submatrices. In [36], another equivalent form of computation is presented, as discussed below.

For the sake of brevity, we restrict our discussion to 1-D here. Consider two adjacent 8-point DCT blocks as $X_1$ and $X_2$ (column vectors) and their 4-point truncated approximations as $X_1^d$ and $X_2^d$, respectively. Let us represent the $8 \times 8$ DCT matrix $C_8$ as $[T_L \quad T_R]$, where $T_L$ and $T_R$ are respective $8 \times 4$ submatrices. Then the downsampled 8-point DCT block ($Y$) is computed as

$$\begin{aligned} Y &= [T_L \quad T_R] \begin{bmatrix} C_4^T & 0_4 \\ 0_4 & C_4^T \end{bmatrix} \begin{bmatrix} X_1^d \\ X_2^d \end{bmatrix}, \\ &= (T_L C_4^T) X_1^d + (T_R C_4^T) X_2^d, \\ &= (R_d + S_d) X_1^d + (R_d - S_d) X_2^d, \\ &= R_d (X_1^d + X_2^d) + S_d (X_1^d - X_2^d), \end{aligned} \qquad (5.29)$$

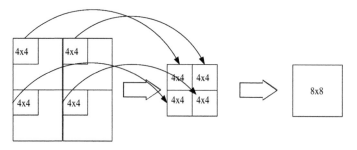

**Figure 5.4**: Image halving [36]: four $4 \times 4$ approximated DCT coefficients of adjacent blocks are merged into an $8 \times 8$ DCT block.

where $R_d$ and $S_d$ are found to be sparse matrices as shown below.

$$
R_d = \begin{bmatrix}
.7071 & 0 & 0 & 0 \\
0 & .296 & 0 & .0162 \\
0 & 0 & 0 & 0 \\
0 & .5594 & 0 & -.069 \\
0 & 0 & .7071 & 0 \\
0 & -.2492 & 0 & .3468 \\
0 & 0 & 0 & 0 \\
0 & .1964 & 0 & .6122
\end{bmatrix},
\tag{5.30}
$$

and,

$$
S_d = \begin{bmatrix}
0 & 0 & 0 & 0 \\
.6407 & 0 & -.0528 & 0 \\
0 & .7071 & 0 & 0 \\
-.2250 & 0 & .3629 & 0 \\
0 & 0 & 0 & 0 \\
.1503 & 0 & .5432 & 0 \\
0 & 0 & 0 & .7071 \\
-.1274 & 0 & -.2654 & 0
\end{bmatrix}.
\tag{5.31}
$$

In 2-D, the equivalent form of Eq. (5.27) is given as

$$
Y = (U + V)R_d^T + (U - V)S_d^T,
\tag{5.32}
$$

where

$$
\begin{aligned}
U &= R_d(X_{00}^d + X_{10}^d) + S_d(X_{00}^d - X_{10}^d), \\
V &= R_d(X_{01}^d + X_{11}^d) + S_d(X_{01}^d - X_{11}^d).
\end{aligned}
\tag{5.33}
$$

It is shown in [36] that the above computation is performed with 1.25 multiplications and 1.25 additions per pixel of the original image.

As mentioned before, the algorithm presented above essentially performs subband approximation before composition of smaller blocks to the larger one. These operations may be performed in the reverse order also. In this case, four

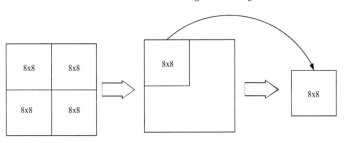

**Figure 5.5**: Image halving [103]: four $8 \times 8$ are composed and the composed block is approximated to an $8 \times 8$ DCT block.

(a)          (b)

**Figure 5.6**: Image halving: (a) IHAC (37.43 dB with respect to the image in Figure 5.2(b) with JPQM as 9.14) and (b) IHCA (36 dB with respect to the image in Figure 5.2(b) with JPQM as 9.31).

$8 \times 8$ blocks are composed into a $16 \times 16$ block and the latter is decimated to an $8 \times 8$ one through subband approximation (see Figure 5.5). This approach is referred to as *Image Halving through block Composition followed by subband Approximation* (IHCA) [103]. Suppose $X_{ij}, 0 \leq i, j \leq 1$ are four adjacent $8 \times 8$ blocks. In that case, computations related to the IHCA technique are given below.

$$Z = A_{(2,8)} \begin{bmatrix} X_{00} & X_{01} \\ X_{10} & X_{11} \end{bmatrix} A_{(2,8)}^T. \tag{5.34}$$

$$Y = \frac{1}{2} [Z(k,l)] \, 0 \leq k,l \leq 7. \tag{5.35}$$

However, in [103], it is observed that the IHCA is computationally more expensive than the former technique, requiring 6 multiplications and 5.75 additions per-pixel of the original image.

### 5.1.3.2   Image Doubling

For image doubling operation, we follow the computational steps as depicted in Figure 5.7. In this case, an $8 \times 8$ DCT block (say $X$) is decomposed into four $4 \times 4$ DCT blocks in the following way:

$$\begin{bmatrix} Y_{00}^d & Y_{01}^d \\ Y_{10}^d & Y_{11}^d \end{bmatrix} = A_{(2,4)}^T X A_{(2,4)}. \tag{5.36}$$

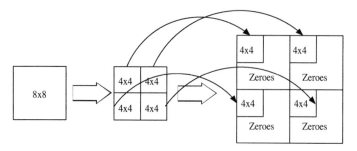

**Figure 5.7**: Image doubling [36]: an $8 \times 8$ DCT block is decomposed into four $4 \times 4$ DCT blocks, each of which is approximated to an $8 \times 8$ DCT block with zero-padding.

Each of the $4 \times 4$ blocks are then approximated to an $8 \times 8$ block by proper scaling (multiplying by 2) and zero-padding. For example, from $Y_{00}^d$ the following upsampled block $Y_{00}$ is generated:

$$Y_{00} = 2 \begin{bmatrix} Y_{00}^d & 0_4 \\ 0_4 & 0_4 \end{bmatrix}. \tag{5.37}$$

As the above technique performs decomposition before the subband approximation, it is referred to as *Image Doubling through Decomposition followed by Approximation* (IDDA) [103].

In [36], similar to downsampling, the above computation is performed efficiently with the help of $R_d$ and $S_d$ (refer to Eq. (5.33)) with 1.25 multiplications and 1.25 additions per-pixel of the upsampled image.

In another approach [106], at first, subband approximation is carried out to an $8 \times 8$ block for converting it into a $16 \times 16$ block. Then the larger block is decomposed into four adjacent $8 \times 8$ blocks (refer to Figure 5.8). This technique is referred to as *Image Doubling through Approximation followed by Decomposition* (IDAD). It has been reported in [106] that this approach provides better quality of results than the former technique. However, the former approach is more computationally efficient.

In [106], how the latter approach (that is, IDAD) could be efficiently implemented for the image doubling operation is discussed. Let $B$ be a block of DCT coefficients in the original image. Let $\tilde{B}$ be the approximated $16 \times 16$ DCT coefficients obtained from $B$ as

$$\tilde{B} = \begin{bmatrix} \hat{B} & 0_8 \\ 0_8 & 0_8 \end{bmatrix}. \tag{5.38}$$

In Eq. (5.38), $\hat{B}$ is obtained from $B$ according to Eq. (2.165) or Eq. (2.166) of Chapter 2. As before, we denote $C_{16}$ and $C_8$ as $16 \times 16$ and $8 \times 8$ DCT matrices, respectively. Let us also represent the $C_{16}$ matrix by its four $8 \times 8$

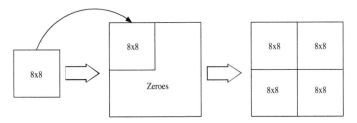

**Figure 5.8**: Image doubling [106]: an $8 \times 8$ DCT block is approximated to a $16 \times 16$ DCT block with zero-padding, which is subsequently decomposed into four $8 \times 8$ DCT blocks.

submatrices as follows:

$$C_{16} = \begin{bmatrix} T_{LL} & T_{LH} \\ T_{HL} & T_{HH}. \end{bmatrix}. \tag{5.39}$$

Hence, the four $8 \times 8$ DCT blocks, $B_{ij}, 0 \leq i, j \leq 1$, could be computed by the following set of equations:

$$\begin{aligned} B_{00} &= P_u^T \hat{B} P_u^T, \\ B_{12} &= P_u^T \hat{B} Q_u^T, \\ B_{21} &= R_u^T \hat{B} P_u^T, \\ B_{22} &= R_u^T \hat{B} Q_u^T. \end{aligned} \tag{5.40}$$

where $P_u = T_{LL} C_8^T$, $Q_u = T_{LH} C_8^T$, and $R_u = T_{HL} C_8^T$. We observe that the conversion matrices, $P_u$, $Q_u$, and $R_u$, are all $8 \times 8$ matrices, and they are sparse. Computations could be further reduced by considering $P_u = E + F$ and $Q_u = E - F$, where $E$ and $F$ are defined as

$$E = \begin{bmatrix} .7071 & 0 & 0 & 0 & 0 & 0 & 0 & 0 \\ 0 & .2986 & 0 & 0.0241 & 0 & 0.0071 & 0 & 0.0018 \\ 0 & 0 & 0 & 0 & 0 & 0 & 0 & 0 \\ 0 & .5446 & 0 & -.0951 & 0 & -0.0235 & 0 & -0.0057 \\ 0 & 0 & .7071 & 0 & 0 & 0 & 0 & 0 \\ 0 & -.2219 & 0 & .4008 & 0 & .0493 & 0 & 0.011 \\ 0 & 0 & 0 & 0 & 0 & 0 & 0 & 0 \\ 0 & .1509 & 0 & .4971 & 0 & -.1078 & 0 & -0.0196 \end{bmatrix}, \tag{5.41}$$

and

Figure 5.9: Image doubling through IDDA (JPQM = 5.26).

$$
F = \begin{bmatrix}
0 & 0 & 0 & 0 & 0 & 0 & 0 & 0 \\
.6376 & 0 & -.0585 & 0 & -.0125 & 0 & -.0039 & 0 \\
0 & .7071 & 0 & 0 & 0 & 0 & 0 & 0 \\
-.2153 & 0 & .3812 & 0 & .0436 & 0 & 0.0128 & 0 \\
0 & 0 & 0 & 0 & 0 & 0 & 0 & 0 \\
.1326 & 0 & .5081 & 0 & -.1061 & 0 & -.0253 & 0 \\
0 & 0 & 0 & 16 & 0 & 0 & 0 & 0 \\
-.0985 & 0 & -.2024 & 0 & .4065 & 0 & .0476 & 0
\end{bmatrix}. \tag{5.42}
$$

In [106] it is shown that these computations are carried out with 4 multiplications and 3.375 additions per pixel of the upsampled image. In this approach, high-frequency components in the upsampled blocks are not exactly zeroes as opposed to the previous one [36]. This makes marginal improvement in the quality of resized images.

## 5.1.4　Performance Analysis

Apart from computational costs, there are other performance measures for comparing the different techniques discussed above. One way of measuring the performance of both the image halving and image doubling techniques in tandem is to get the output generated by them when they are applied one after another to an input image. Then the result is compared with the

Figure 5.10: Image doubling through IDAD ( JPQM = 6.88).

**Table 5.1**: (PSNR in DB, JPQM) values after halving and doubling a grey-level image

| Images | IHCMP-IDCMP | IHAC-IDDA | IHAC-IDAD | IHCA-IDDA | IHCA-IDAD |
|--------|-------------|-----------|-----------|-----------|-----------|
| Lena | (35.36 , 8.67) | (34.7, 7.36) | (35.02 , 8.86) | (34.41 , 6.82) | (35.36 , 8.67) |
| Watch | (30.05, 7.5) | (29.21 , 6.2) | (29.68 , 7.77) | (28.9 , 5.62) | (30.05 , 7.5) |
| Cap | (34.75 , 8.79) | (34.41 , 7.1 ) | (34.44 , 8.9 ) | (34.11 , 6.79) | (34.75 , 8.79) |
| F-16 | (33.22 , 8.76) | (32.48 , 7.32) | (32.86 , 9.0) | ( 32.16 , 6.74) | (33.22 , 8.76) |

original image. In this case, the PSNR measure is chosen to study the faithful reconstruction at the end. In the block DCT space, JPQM is also used to observe the quality of images in terms of occurrences of blocking artifacts in the resulting image. In Table 5.1 we present these measures when a pair of image halving and image doubling techniques are applied to a set of images.[1] Note that the IHCMP-IDCMP techniques and IHCA-IDAD provide identical performance measures for all these images.

---

[1] Obtained from *http://vision.ai.uiuc.edu/~dugad/draft/dct.html*

## 5.2    Resizing with Integral Factors

The techniques for image halving and doubling are extended also for down-sampling or upsampling an image with integral factors. For example, an $L \times M$ downsampling operation reduces the height $(h)$ and width $(w)$ of an image to $\frac{h}{L}$ and $\frac{w}{M}$, respectively. Similarly the upsampling operation converts to an image of size $hL \times wM$. In [62], using the similar application of the multiplication–convolution property, resizing with integral factors has been demonstrated for $L = M = 3$. However, the technique is limited by the fact that both the integral factors are required to be the same (i.e., $L = M$) along the horizontal and vertical directions. Forms and expressions of convolution matrices (see Eqs. (5.18) and (5.24)) also vary widely depending upon the factor $L$. In [103], however, the same computational steps for image halving and doubling are adapted conveniently under a general framework. In these techniques the factors need not be the same for both the directions. In the following subsection, these concepts are elaborated.

### 5.2.1    $L \times M$ Downsampling Algorithm (LMDS)

The LMDS algorithm is based on the strategy of block composition followed by subband approximation similar to the IHCA. In this case, $L \times M$ numbers of $N \times N$ DCT blocks are converted into a block of $LN \times MN$-DCT. Then, from this composed one, the $N \times N$ DCT coefficients of the downsampled block are obtained using the subband relationship. The subband approximation used here is extended from Eq. (2.166) of Chapter 2. This is stated below.

**Theorem 5.1** *Let $X_{LL}(k,l), 0 \le k \le N - 1, 0 \le l \le N - 1$, be the DCT coefficients of an $N \times N$ block of the downsampled image. Then the DCT coefficients of an $LN \times MN$ block are approximated by the following:*

$$X(k,l) = \begin{cases} \sqrt{LM}X_{LL}(k,l), & k,l = 0,1,\ldots,N-1, \\ 0, & otherwise. \end{cases} \qquad (5.43)$$

$\square$

The algorithm for downsampling of images by $L \times M$ is described below.
**Algorithm** $L \times M\_Downsampling(LMDS)$

    For every adjacent $L \times M$ block, $\{ X_{i,j}, \ 0 \le i \le L-1, \ 0 \le j \le M-1 \}$, of the input image, perform the following computations:
    {

(a) Compose an $LN \times MN$ DCT block $Z$ as follows:

$$Z = A_{(L,N)} \begin{bmatrix} X_{0,0} & X_{0,1} & \cdots & X_{0,(M-1)} \\ X_{1,0} & X_{1,1} & \cdots & X_{1,(M-1)} \\ \cdot & \cdot & \cdots & \cdot \\ X_{(L-1),0} & X_{(L-1),1} & \cdots & X_{(L-1),(M-1)} \end{bmatrix} A_{(M,N)}^T.$$

(5.44)

(b) Get the resulting $N \times N$-DCT coefficients of the downsampled image ( $Y$ ) from Eq. (5.43).

$$Y = \sqrt{\frac{1}{LM}} \left[ Z(k,l) \right]_{0 \le k,l \le N-1}.$$

(5.45)

}

End $L \times M\_Downsampling(LMDS)$

For efficient computation, we may not compute the full $Z$ matrix in the above algorithm. It is sufficient to compute only the topmost and leftmost $N \times N$ submatrix of $Z$. Hence, $Y$ could be computed effectively by the following expression:

$$Y = \sqrt{\frac{1}{LM}} (A_{00}^{(L,N)} (B_{00}(A_{00}^{(M,N)})^T + B_{01}(A_{01}^{(M,N)})^T) + A_{01}^{(L,N)}(B_{10}(A_{00}^{(M,N)})^T + B_{11}(A_{01}^{(M,N)})^T)),$$

(5.46)

where $A_{00}^{(L,N)}$ (of size $N \times N$) and $A_{01}^{(L,N)}$ (of size $N \times (L-1)N$) are submatrices of $A_{(L,N)}$. Similarly, $A_{00}^{(M,N)}$ (of size $N \times N$) and $A_{01}^{(M,N)}$ (of size $N \times (M-1)N$) are submatrices of $A_{(M,N)}$. Again $B_{00}$, $B_{01}$, $B_{10}$, and $B_{11}$ are submatrices of the $L \times M$ DCT blocks, and their sizes are $N \times N$, $N \times (M-1)N$, $(L-1)N \times N$, and $(L-1)N \times (M-1)N$, respectively. Hence following the *general computational complexity model of matrix multiplication* (GCCMMM) (refer to Section 1.7.1 of Chapter 1) the total numbers of multiplications ($n_m^{(LMDS)}$) and additions ($n_a^{(LMDS)}$) for the LMDS algorithm are given by

$$\begin{aligned} n_m^{(LMDS)} = \quad & n_m(A_{00}^{(M,N)}; N, N, N) + n_m(A_{00}^{(M,N)}; N, N, (L-1)N) \\ + \ & n_m(A_{01}^{(M,N)}; N, (M-1)N, N) + n_m(A_{01}^{(M,N)}; N, (M-1)N, (L-1)N) \\ + \ & n_m(A_{00}^{(L,N)}; N, N, N) + n_m(A_{01}^{(L,N)}; N, (L-1)N, N) \\ + \ & N^2. \end{aligned}$$

(5.47)

$$\begin{aligned} n_a^{(LMDS)} = \quad & n_a(A_{00}^{(M,N)}; N, N, N) + n_a(A_{00}^{(M,N)}; N, N, (L-1)N) \\ + \ & n_a(A_{01}^{(M,N)}; N, (M-1)N, N) + n_a(A_{01}^{(M,N)}; N, (M-1)N, (L-1)N) \\ + \ & n_a(A_{00}^{(L,N)}; N, N, N) + n_a(A_{01}^{(L,N)}; N, (L-1)N, N) \\ + \ & 3N^2. \end{aligned}$$

(5.48)

In this computation, $N^2$ multiplications are required for subband approximation for deriving the subsampled DCT coefficients (refer to Eq. (5.43)). In Eq. (5.47), this is also taken care of. In Table 5.2, the computational cost for the LMDS with different values of $L$ and $M$ are provided [103]. In this table, $n_m$ and $n_a$ denote, respectively, the number of multiplications and additions required for processing a pixel of the original image.

Table 5.2: Per-pixel (of the original image) computational cost of LMDS

| $L \times M$ | LMDS $(n_m, n_a)$ | $L \times M$ | LMDS $(n_m, n_a)$ |
|---|---|---|---|
| $1 \times 2$ | $(5.5, 4.5)$ | $2 \times 2$ | $(7,6)$ |
| $1 \times 3$ | $(5.08, 4.54)$ | $3 \times 3$ | $(6, 5.5)$ |
| $2 \times 3$ | $(6.08, 5.54)$ | $4 \times 4$ | $(5.22, 7.38)$ |
| $2 \times 5$ | $(5.3, 6.25)$ | $5 \times 5$ | $(5.2, 6.42)$ |
| $2 \times 7$ | $(4.57, 6.25)$ | $6 \times 6$ | $(3.82, 6.60)$ |
| $3 \times 4$ | $(5.31, 6.97)$ | $7 \times 7$ | $(4.43, 6.39)$ |
| $3 \times 5$ | $(5.25, 6.23)$ | $8 \times 8$ | $(3.95, 7.78)$ |

## 5.2.2   $L \times M$ upsampling Algorithm (LMUS)

The LMUS technique is an extension of the IDAD algorithm (refer to Section 5.1.3.2), where subband approximation is followed by block decomposition. Here, the approximation is carried out using the same Eq. (5.43). Hence, an $N \times N$ DCT block $B$ of an image is converted into an $LN \times MN$ DCT block $\hat{B}$ of the upsampled image as follows:

$$\hat{B} = \begin{bmatrix} \sqrt{LM}B & 0_{(N,(M-1)N)} \\ 0_{((L-1)N,N)} & 0_{((L-1)N,(M-1)N)} \end{bmatrix}, \tag{5.49}$$

where $0_{(a,b)}$ denotes a matrix of $a \times b$ zero elements. After conversion, DCT-block decomposition is used to obtain $L \times M$ numbers of $N \times N$-DCT blocks in the upsampled image. This algorithm is referred to as $L \times M$-*UpSampling* (LMUS).

In the LMUS, for efficient computation, full decomposition is not carried out. Consider that $IA_{(L,N)}$ and $IA_{(M,N)}$ represent $A_{(L,N)}^{-1}$ and $A_{(M,N)}^{-1}$, respectively. Their submatrices are also defined in the same way that submatrices of $A_{(L,N)}$ (or $A_{(M,N)}$) are defined in the computation of the LMDS (see the previous section). The resulting decomposition operation is factored into the following forms:

$$\begin{bmatrix} IA_{00}^{(L,N)} & IA_{01}^{(L,N)} \\ IA_{10}^{(L,N)} & IA_{11}^{(L,N)} \end{bmatrix} \begin{bmatrix} \hat{B} & 0_{(N,(M-1)N)} \\ 0_{((L-1)N,N)} & 0_{((L-1)N,(M-1).N)} \end{bmatrix} \begin{bmatrix} (IA_{00}^{(M,N)})^T & (IA_{10}^{(M,N)})^T \\ (IA_{01}^{(M,N)})^T & (IA_{11}^{(M,N)})^T \end{bmatrix}$$

$$= \begin{bmatrix} IA_{00}^{(L,N)}\hat{B}(IA_{00}^{(M,N)})^T & IA_{00}^{(L,N)}\hat{B}(IA_{10}^{(M,N)})^T \\ IA_{01}^{(L,N)}\hat{B}(IA_{00}^{(M,N)})^T & IA_{10}^{(L,N)}\hat{B}(IA_{10}^{(M,N)})^T \end{bmatrix}. \tag{5.50}$$

The numbers of multiplications and additions are given by the following equations:

$$\begin{aligned} n_m^{(LMUS)} = \quad & n_m(IA_{00}^{(L,N)}; N, N, N) + n_m(IA_{10}^{(L,N)}; (L-1)N, N, N) \\ + \quad & n_m(IA_{00}^{(M,N)}; N, N, N) + n_m(IA_{10}^{(M,N)}; N, (M-1)N, N) \\ + \quad & n_m(IA_{00}^{(M,N)}; N, N, (L-1)N) + n_m(IA_{10}^{(M,N)}; (M-1)N, N, (L-1)N) \\ + \quad & N^2, \end{aligned} \tag{5.51}$$

Table 5.3: Per-pixel computational cost (of the upsampled image) of LMUS

| $L \times M$ | LMUS $(n_m, n_a)$ | $L \times M$ | LMUS $(n_m, n_a)$ |
|---|---|---|---|
| $1 \times 2$ | (5.5,3.5) | $2 \times 2$ | (7,5.25) |
| $1 \times 3$ | (5.2,3.54) | $3 \times 3$ | (6.16,4.72) |
| $2 \times 3$ | (6.21,4.71) | $4 \times 4$ | (7.88,6.56) |
| $2 \times 5$ | (6.65,5.35) | $5 \times 5$ | (6.82,5.58) |
| $2 \times 7$ | (6.5,5.32) | $6 \times 6$ | (6.93,5.74) |
| $3 \times 4$ | (7.47,6.14) | $7 \times 7$ | (6.63,5.51) |
| $3 \times 5$ | (6.63,5.36) | $8 \times 8$ | (7.93,6.89) |

$$
\begin{aligned}
n_a^{(LMUS)} = \ & n_a(IA_{00}^{(L,N)}; N, N, N) + n_a(IA_{10}^{(L,N)}; (L-1)N, N, N) \\
+ \ & n_a(IA_{00}^{(M,N)}; N, N, N) + n_a(IA_{10}^{(M,N)}; N, (M-1)N, N) \\
+ \ & n_a(IA_{00}^{(M,N)}; N, N, (L-1)N) + n_a(IA_{10}^{(M,N)}; (M-1)N, N, (L-1)N).
\end{aligned}
\tag{5.52}
$$

The cost of computing a pixel of the upsampled image is shown in Table 5.3 [103].

Typical examples of image downsampling and upsampling operations with a factor of $3 \times 2$ are shown in Figure 5.11 (a) and (b), respectively.

## 5.3 Resizing with Arbitrary Factors

In [134], a downsampling algorithm for resizing with any arbitrary factor $R_h \times R_w$ is presented, where for every $8 \times 8$ block of the downsampled image, its corresponding area of influence in the original image is considered. Pixels related to that area of influence lie in a set of adjacent blocks. They are extracted to form the block of size $8R_h \times 8R_w$. Finally, the DCT of this block is approximated to the $8 \times 8$ one. All these operations are carried out in the transform domain using the linear and distributive property of DCT, and the computation is aided by using a lookup table for extraction of pixels under different overlapping conditions of the area of influence with an $8 \times 8$ block in the original image. Later, in [112] the concept is extended to take care of upsampling also. Though these techniques are applicable for any arbitrary factor, computation of area of influence and subsequent extraction of pixels in overlapping zones in a block varies significantly depending on the values of resizing factors.

In some other approaches, downsampling and upsampling of images with integral factors are used in tandem to perform resizing with any arbitrary

(a)                                                                                (b)

**Figure 5.11**: Image resizing by a factor of $3 \times 2$: (a) downsampled image (JPQM = 9.19) and (b) upsampled image (JPQM = 5.86).

rational factors in the form of $\frac{P}{Q} \times \frac{R}{S}$, where $P$, $Q$, $R$, and $S$ are positive integers. In [62], resizing by $\frac{P}{Q} \times \frac{P}{Q}$ is performed by first performing upsampling with $P \times P$ and then by downsampling the upsampled image by $Q \times Q$. These two cascading tasks may be interchanged also to provide the resized image with the same factor. Though the computation in the latter approach is significantly less compared to the former one, the quality of resized images greatly suffers in this case.

In the same way, the LMDS and the LMUS algorithms operate in tandem to perform resizing with factors of rational numbers. In this case, the resizing factors along two different directions need not be same. If the LMUS is followed by the LMDS, upsampling is first carried out on the image with the integral factor of $P \times R$, and then the downsampling operation with the factor of $Q \times S$ is applied. Alternatively, the downsampling operation (by a factor of $Q \times S$) may be applied before upsampling (with $P \times R$ factor). The former approach is referred here as *UD-Resizing Algorithm* (UDRA), and the latter is called *DU-Resizing Algorithm* (DURA). The latter approach speeds up the computation. In Figure 5.12, a typical example of conversion of an HDTV frame (of size $1080 \times 1920$) to an NTSC one (of size $480 \times 640$) is shown [103]. For this conversion, the DURA takes $6.76M + 0.77A$ operations per-pixel (of the original image), whereas the UDRA takes $182.625M + 21A$ operations per-pixel (of the original image). Though the latter task involves a significantly higher number of operations, the quality of resized images is much improved.

An almost similar strategy is followed in [125]. However, instead of identifying the steps as downsampling or upsampling, in this case, an $8 \times 8$ block of the image is resized to a block of size $P \times P$ by applying $P$-point IDCT over the approximated coefficients, where $P$ may be greater or less than 8. When $P$ is greater than 8, the DCT block is zero-padded, otherwise it is truncated. In both cases, appropriate scaling of the coefficients is carried out as it is done for subband approximation. This scales the image by the factor of $\frac{P}{8} \times \frac{P}{8}$. In the next stage, $Q$-point DCT is applied to the intermediate image and converts each $Q \times Q$ DCT block to an $8 \times 8$ one using subband approximation. This is achieved by scaling the intermediate image by the factor of $\frac{8}{Q} \times \frac{8}{Q}$. Hence, the effective resizing factor becomes $\frac{P}{Q} \times \frac{P}{Q}$. The computation depends upon the values of $P$ and $Q$. By keeping $\frac{P}{Q}$ constant, we may vary the values of $P$ and $Q$ to achieve scalable performance of resizing leading to a trade-off between the computational cost and quality of resized image.

The technique presented in [125] is, however, limited by the fact that it is applicable for uniform resizing factors along both directions. However, a little modification of this technique will make it work with nonuniform factors. Moreover, in the same computational stage, rational resizing takes place. This makes the basic resizing operation more general, other than the conventional downsampling and upsampling operations. For example, for resizing with a factor $\frac{P}{Q} \times \frac{R}{S}$, if $\frac{P}{Q} > 1$ and $\frac{R}{S} < 1$, we cannot refer to this operation as

**Figure 5.12**: Conversion of an HDTV frame ($1080 \times 1920$) to an NTSC frame ($480 \times 640$): (a) HDTV, (b) NTSC by UDRA (JPQM = 10.21), (c) NTSC by DURA (JPQM = 8.35), and (d) NTSC by HRA (JPQM = 9.92).

downsampling or upsampling. In that case, we refer to this operation as *hybrid resizing*. In the next section, we elaborate on this concept.

---

## 5.4   Hybrid Resizing

The *hybrid resizing algorithm* (HRA) follows the similar computational sequence of block composition, subband approximation and block decomposition. However, it distinctly differs in the way subband approximation is carried out. In this case, the subband approximation relationship is put into a more general form as given below.

Let $\mathbf{X}$ be a DCT block of size $QN \times SN$. The subband approximated DCT block $\mathbf{Y}$ of size $PN \times RN$ is given by the following expression.

$$\mathbf{Y} = \sqrt{\frac{PR}{QS}}\hat{\mathbf{X}}, \qquad (5.53)$$

where $\hat{\mathbf{X}}$ are obtained by truncating or zero-padding of the coefficients of $\mathbf{X}$ in the respective dimensions depending upon the ordered relationship between $P$ and $Q$, and $R$ and $S$, as elaborated later.

Using the above relationship, in the HRA technique, $Q \times S$ adjacent DCT blocks are converted to $P \times R$ DCT blocks. The algorithm for this computation is described below.

**Algorithm** *Hybrid_Resizing_Algorithm* (HRA)
**Input**: $Q \times S$ adjacent DCT blocks, each of size $N \times N$.
**Output**: $P \times R$ adjacent DCT blocks, each of size $N \times N$.

1. **Block composition:** In this step, $Q \times S$ input blocks are composed into a block of size $QN \times SN$.

2. **Subband approximation:** Let $\mathbf{Y}$ denote the $PN \times RN$ subband approximated block obtained from the input block $\mathbf{X}$ of size $QN \times SN$. Then the relationship between $\mathbf{Y}$ and $\mathbf{X}$ is expressed in the following equation:

$$
\begin{aligned}
\mathbf{Y} &= \sqrt{\frac{PR}{QS}}\,[\mathbf{X}]_{0\le i<PN,0\le j<RN}, & P \le Q, R \le S, \\[2mm]
&= \sqrt{\frac{PR}{QS}}\begin{bmatrix} [\mathbf{X}]_{0\le i<QN,0\le j<RN} \\ 0_{(P-Q)N\times RN} \end{bmatrix}, & P > Q, R \le S, \\[2mm]
&= \sqrt{\frac{PR}{QS}}\begin{bmatrix} [\mathbf{X}]_{0\le i<PN,0\le j<SN} & 0_{PN\times(R-S)N} \end{bmatrix}, & P \le Q, R > S, \\[2mm]
&= \sqrt{\frac{PR}{QS}}\begin{bmatrix} [\mathbf{X}]_{0\le i<QN,0\le j<SN} & 0_{QN\times(R-S)N} \\ 0_{(P-Q)N\times SN} & 0_{(P-Q)N\times(R-S)N} \end{bmatrix}, & P > Q, R > S.
\end{aligned}
$$

$$(5.54)$$

Table 5.4: JPQM values of resized images of *Watch*

| Resizing factor | JPQM | | |
|---|---|---|---|
| | HRA | DURA | UDRA |
| $\frac{2}{3} \times \frac{4}{5}$ | 9.01 | 6.58 | 9.37 |
| $\frac{3}{5} \times \frac{5}{4}$ | 8.74 | 6.15 | 8.86 |
| $\frac{2}{3} \times \frac{3}{2}$ | 8.55 | 6.90 | 8.90 |
| $\frac{3}{5} \times \frac{2}{3}$ | 8.80 | 7.12 | 9.12 |

3. **Block decomposition:** In this step, $\mathbf{Y}$ is decomposed into $P \times R$ blocks, each of size $N \times N$.

End *Hybrid_Resizing_Algorithm*

We demonstrate here typical cases of different types of resizing for the image *Watch* in Figure 5.13. In Table 5.4, we provide JPQM values for different techniques.

## 5.4.1    Computational Cost

As the subband approximation is carried out in a more general framework, it is difficult to design efficient techniques by computing only the relevant coefficients of the resulting DCT blocks. In this case, full composition and decomposition are carried out. However, exploiting the relationships among the elements of the composition matrix and using its sparsity, the computational cost is significantly reduced (refer to properties of block composition matrices and number of operations required to perform composition and decomposition in Section 2.2.4.5 of Chapter 2). In Table 5.5, we provide the computational costs for HRA, DURA, and UDRA. All these costs are expressed as the numbers of multiplications ($n_m$) and additions ($n_a$) per-pixel of the original image. The usefulness of the HRA lies in the fact that it significantly reduces the number of multiplications compared to the UDRA, though the quality of the resized image (refer to Table 5.4) is almost at par with those obtained from the latter. This is demonstrated also by using the HRA to convert the HDTV frame to the NTSC frame as shown in Figure 5.12(d). The JPQM value of the resized image is quite close to the one obtained by the UDRA. In this case, the HRA takes 42.5 multiplications and 75.15 additions for processing a pixel of the original image. The cost is significantly reduced compared to the UDRA. The DURA is the fastest, though images resized by this technique suffer heavily from the appearance of blocking artifacts (see Figure 5.12(c)). This fact is also supported by low JPQM values in Table 5.5.

Table 5.5: Per-pixel computational cost of arbitrary resizing algorithms

| Resizing factor | $(n_m, n_a)$ | | |
|---|---|---|---|
| | HRA | DURA | UDRA |
| $\frac{2}{3} \times \frac{4}{5}$ | (33.6, 53.47) | (9.25, 3.27) | (138.75, 49) |
| $\frac{2}{3} \times \frac{5}{4}$ | (67.15, 91) | (17.8, 10.05) | (142.38, 80.38) |
| $\frac{2}{3} \times \frac{3}{2}$ | (27.58, 40.17) | (13.08, 4.71) | (78.5, 28.25) |
| $\frac{2}{3} \times \frac{2}{3}$ | (27.58, 40.17) | (13.02, 5.27) | (78.13, 31.63) |

## 5.5 Summary

Image resizing in the compressed domain has drawn considerable attention from researchers in recent years. Initially, efforts were mostly concentrated on development of algorithms for image halving and doubling techniques. Later, those approaches are extended for resizing with arbitrary factors. In one such computational approach, the resizing operation is performed as a combination of block composition (or decomposition) and subband approximation of DCT coefficients. Resizing with rational factors is usually performed in a cascading combination of downsampling and upsampling or the reverse. However, using a more general form of subband approximation, we perform this operation in a single step. This type of resizing is referred to here as hybrid resizing as in this case image dimensions are independently up-sized or down-sized.

(a) (b)

(c) (d)

**Figure 5.13**: Hybrid resizing: (a)by $\frac{2}{3} \times \frac{4}{5}$, (b) by $\frac{3}{2} \times \frac{5}{4}$ (c) by $\frac{2}{3} \times \frac{3}{2}$, and (d) by $\frac{3}{2} \times \frac{2}{3}$.

# Chapter 6

## Transcoding

Transcoding of images and videos accomplishes the task of converting them from one format to another. For an image encoded and compressed in a format, the conversion may involve repackaging it into a different compression standard. Image resizing, discussed in the previous chapter, is also a transcoding operation. For a video, there could be various types of transcoding operations, such as transforming it into a different compression standard. Even within the same standard, alteration of its bit rate, frame rate, frame size, etc., are considered transcoding operations. Transcoding is required for different multimedia applications involving their presentation and communication over heterogeneous devices and networks. For example, the resolution of a display device may vary from $240 \times 160$ of a palmtop to as high as $3280 \times 2048$ of an advanced monitor for medical diagnostics. To adapt the display of an image or video of a given resolution, we require to bring it within the permissible display resolution of the device. In most applications, services available for display and communication of video and images are restricted by their adherence to a few selected standards. Hence, transcoding provides an interface for accommodating other representations, not supported by those services. Consider a scenario, in which images are saved in a server in the JPEG2000 format, as it is usually more efficient than the JPEG standard in storage requirement. However, even today, JPEG is the standard of image representation widely supported by various applications for the purpose of display and processing. Many of them may not support JPEG2000. In this case, we have to transcode a JPEG2000 image into a JPEG standard, before feeding it to such an application.

In this chapter, we discuss various issues of transcoding. As most of the image and video standards use the DCT, ICT, or DWT for their representations, first we review techniques for intertransform conversion, directly in their domains. Next we discuss a few specific cases of transcoding. Image resizing was discussed in the previous chapter. In this chapter, we also consider the video resizing problem.

## 6.1    Intertransforms Conversion

As in image compression standards such as JPEG and JPEG2000, the DCT and DWT are used, in transcoding operations for converting one form to the other, intertransform exchange between these two is needed. Similarly, for videos, the DCT and ICT are common transforms used in compression standards such as MPEG-2 and H.264, respectively. In this case, conversion between DCT and ICT is required during transcoding of a video. In this section, we discuss the techniques for performing these operations in the transform domain.

## 6.1.1 DWT to DCT

The DWT is discussed in Chapter 2. However, for the convenience of the reader, the block diagram of the DWT and its inverse is reillustrated in Figure 6.1. From the diagram, we observe that transcoding the DWT coefficients into DCTs involves synthesis of these coefficients. Hence, the computation requires two basic operations, namely:

1. Upsampling of DWT coefficients by inserting zeroes appropriately.

2. Filtering with synthesis filters (say, $h'(n)$ and $g'(n)$ of Figure 6.1).

In the following subsections, we discuss how the above computations are performed efficiently and directly in the transform domain. For the convenience of discussion, initially we restrict ourselves to the 1-D.

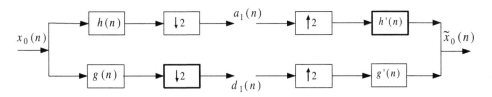

**Figure 6.1**: Forward and inverse DWT in 1-D. (From [147] with permission from the author.)

### 6.1.1.1 Inserting Zeroes in DWT Coeffcients

Following theorems state how the DCTs of zero-inserted sample values are computed. Zeroes are inserted in alternate sample positions of $a_1(n)$ and $d_1(n)$ (refer to Figure 6.1). When zeroes are inserted in odd sample locations of the upsampled sequence (refer to Eq. (2.146) of Chapter 2), the upsampling is referred to as *even upsampling*. As shown in Figure 6.1, $a_1(n)$ goes through even upsampling. On the other hand, for $d_1(n)$, zeroes are inserted before every sample location so that every even location of the upsampled sequence has a zero (refer to Eq. (2.147) of Chapter 2). The latter is referred to as *odd upsampling*.

**Theorem 6.1** *Let $x_{ue}(m)$, $m = 0, 1, 2, ..., 2N-1$ be obtained from $x(n)$, $n = 0, 1, 2, ..., N-1$ by inserting zeroes after every sample (refer to Eq. (2.146) of chapter 2). The DCT of $x_{ue}(m)$, is given by the following expression:*

$$X_{ue}(k) = \sqrt{\frac{1}{N}}\alpha(k) \sum_{n=0}^{N-1} x(n) \cos\left(\frac{\pi(n + \frac{1}{4})k}{N}\right), \quad 0 \leq k \leq 2N - 1. \quad (6.1)$$

*where $\alpha(k)$ is defined in Eq. (2.89) in Chapter 2.*

**Proof:**

The DCT of $\check{x}_{ue}(m)$ is defined as given below (refer Eq. (2.75) of Chapter 2).

$$C_{2e}\{x_{ue}(m)\} = \sqrt{\frac{2}{2N}}\alpha(k) \sum_{m=0}^{2N-1} x_{ue}(m) \cos\left(\frac{\pi(m+\frac{1}{2})k}{2N}\right), \; 0 \le k \le 2N-1.$$

By replacing the variable $m$ with $2n$ in the above equation, the resulting expression becomes

$$X_{ue}(k) = \sqrt{\frac{1}{N}}\alpha(k) \sum_{2n=0}^{2N-1} x(2n) \cos\left(\frac{\pi(2n+\frac{1}{2})k}{2N}\right), \; 0 \le k \le 2N-1.$$

From the above, we obtain the expression as stated in the theorem:

$$X_{ue}(k) = \sqrt{\frac{1}{N}}\alpha(k) \sum_{n=0}^{N-1} x(n) \cos\left(\frac{\pi(n+\frac{1}{4})k}{N}\right), \; 0 \le k \le 2N-1.$$

$\square$

**Theorem 6.2** *Let* $x_{uo}(m)$, $m = 0, 1, 2, ..., 2N-1$ *be obtained by inserting zeroes before every sample of* $x(n)$, $n = 0, 1, 2, ..., N-1$ *(refer to Eq. (2.147) of chapter 2). The DCT of* $x_{uo}(m)$ *is given by the following expression:*

$$X_{uo}(k) = \sqrt{\frac{1}{N}}\alpha(k) \sum_{n=0}^{N-1} x(n) \cos\left(\frac{\pi(n+\frac{3}{4})k}{N}\right), \; 0 \le k \le 2N-1. \quad (6.2)$$

*where* $\alpha(k)$ *is defined in Eq. (2.89) in Chapter 2.*

**Proof:**

The DCT of $x_{uo}(m)$ is computed by the following expression (refer to Eq. (2.75) of Chapter 2):

$$C_{2e}\{x_{uo}(m)\} = \sqrt{\frac{2}{2N}}\alpha(k) \sum_{m=0}^{2N-1} x_{uo}(m) \cos\left(\frac{\pi(m+\frac{1}{2})k}{2N}\right), \; 0 \le k \le 2N-1.$$

Let us replace the variable $m$ with $2n+1$ in the above expression and obtain the following.

$$X_{uo}(k) = \sqrt{\frac{1}{N}}\alpha(k) \sum_{n=0}^{N-1} x(2n+1) \cos\left(\frac{\pi(2n+1+\frac{1}{2})k}{2N}\right), \; 0 \le k \le 2N-1.$$

From the above, the expression is given as

$$X_{uo}(k) = \sqrt{\frac{1}{N}}\alpha(k) \sum_{n=0}^{N-1} x(n) \cos\left(\frac{\pi(n+\frac{3}{4})k}{N}\right), \; 0 \le k \le 2N-1.$$

□

Let us denote the upsampled DCT basis matrices in Eqs. (6.1) and (6.2) as $B_{ue}^{(2N,N)}$ and $B_{uo}^{(2N,N)}$, respectively. Let $C_{2N}$ be the $2N$-point DCT matrix. We observe that the columns of basis matrices $B_{ue}$ and $B_{uo}$ are the even indexed and odd indexed columns of the $2N$-point DCT matrix $C_{2N}$, respectively. For example, to compute 8-point upsampled DCT block from an input sequence of length 4 , we use the conversion matrices $B_{ue}^{(8,4)}$ and $B_{uo}^{(8,4)}$ (Eqs. (6.1) and (6.2)), which are given as follows:

$$
B_{ue}^{(8,4)} = \begin{bmatrix}
0.3536 & 0.3536 & 0.3536 & 0.3536 \\
0.4904 & 0.2778 & -0.0975 & -0.4157 \\
0.4619 & -0.1913 & -0.4619 & 0.1913 \\
0.4157 & -0.4904 & 0.2778 & 0.0975 \\
0.3536 & -0.3536 & 0.3536 & -0.3536 \\
0.2778 & 0.0975 & -0.4157 & 0.4904 \\
0.1913 & 0.4619 & -0.1913 & -0.4619 \\
0.0975 & 0.4157 & 0.4904 & 0.2778
\end{bmatrix}, \tag{6.3}
$$

$$
B_{uo}^{(8,4)} = \begin{bmatrix}
0.3536 & 0.3536 & 0.3536 & 0.3536 \\
0.4157 & 0.0975 & -0.2778 & -0.4904 \\
0.1913 & -0.4619 & -0.1913 & 0.4619 \\
-0.0975 & -0.2778 & 0.4904 & -0.4157 \\
-0.3536 & 0.3536 & -0.3536 & 0.3536 \\
-0.4904 & 0.4157 & -0.0975 & -0.2778 \\
-0.4619 & -0.1913 & 0.4619 & 0.1913 \\
-0.2778 & -0.4904 & -0.4157 & -0.0975
\end{bmatrix}. \tag{6.4}
$$

Given the input as a column vector $\mathbf{x}$ of length 4, the corresponding *even upsampled* and *odd upsampled* DCTs, are given by Eqs. (6.5) and (6.6) respectively.

$$
\mathbf{X_{ue}} = B_{ue}^{(8,4)}\mathbf{x}, \tag{6.5}
$$

$$
\mathbf{X_{uo}} = B_{uo}^{(8,4)}\mathbf{x}, \tag{6.6}
$$

As the upsampled block DCT is computed directly from the wavelet subbands using Eqs. (6.5) and (6.6), it requires fewer operations. These blocks become the input to the synthesis filters, which perform the computation in the transform domain as discussed in Chapter 3. Let us refer to this kind of upsampling with zero insertion in the DCT domain as the *DCT domain upsampling*, and also the DCT obtained by this process as the *upsampled DCT*.

### 6.1.1.2  DCT Domain Upsampling in 2D

In 2-D, the above operations for upsampling (refer to Eqs. (6.5) and (6.6)) are performed on input columns and then on rows of the resulting blocks. Consider $\mathbf{w}$ to be a $4 \times 4$ wavelet coefficient block. Its upsampled DCT block,

$W_{ee}(k,l)$, $0 \leq k, l \leq 7$ (represented as $\mathbf{W_{ee}}$ in matrix form) for LL subband synthesis is computed as

$$\mathbf{W_{ee}} = B_{ue}^{(8,4)} \mathbf{w} B_{ue}^{(8,4)^T}. \tag{6.7}$$

Similarly, $W_{oo}(k,l)$, $0 \leq k, l \leq 7$ for HH subband synthesis is computed as

$$\mathbf{W_{oo}} = B_{uo}^{(8,4)} \mathbf{w} B_{uo}^{(8,4)^T}. \tag{6.8}$$

### 6.1.1.3   Upsampling the DCT for Multilevel DWT

By using the $N$-point DCT matrix $C_N$, we also compute the upsampled DCT directly from the DCT coefficients of the input. In this case, the composite matrix $(B_{uc}^{(2N,N)})$ has the following form:

$$B_{uc}^{(2N,N)} = B_{ue}^{(2N,N)} C_N^T. \tag{6.9}$$

An example of this matrix for $N = 4$ is given below.

$$B_{uc}^{(8,4)} = \begin{bmatrix} 0.7071 & 0 & 0 & 0 \\ 0.1274 & 0.6935 & -0.0528 & 0 \\ 0 & 0.2500 & 0.6533 & -0.1036 \\ 0.1503 & 0 & 0.3629 & 0.5879 \\ 0 & 0.2706 & 0 & 0.6533 \\ 0.2250 & 0 & 0.5432 & -0.3928 \\ 0 & 0.6036 & -0.2706 & -0.2500 \\ 0.6407 & -0.1379 & -0.2654 & 0 \end{bmatrix}. \tag{6.10}$$

We find that the composite conversion matrix $B_{uc}^{(2N,N)}$ is sparse. Hence, the upsampling operation saves computation by exploiting this fact. In the above computation, Eq. (6.9) computes even upsampling only (refer to Eq. (6.1)).

In multilevel transcoding, processing of subbands into DCT blocks may proceed from the lowest level of resolution following the stages of dyadic decomposition. In that case, for obtaining the next higher-level description in the block DCT space, this composite conversion is required. There only the even upsampling in the DCT space is used, as the synthesis for the next stage takes the output from the previous level as its LL subband coefficients. However, the computation is more efficient than the spatial domain approach, if the block size is kept small. As the block size increases, the upsampled DCT computation becomes expensive due to the increase in the size of the conversion matrix. In this case, the upsampled DCT is computed through spatial domain operations.

### 6.1.1.4   Wavelet Synthesis in the Compressed Domain

In Chapter 3, we discussed how filtering, in particular with symmetric FIR, is performed efficiently in the block DCT domain. The same method is extended

**Table 6.1**: Daubechies 9/7 analysis and synthesis filter banks for lossy compression

| | Analysis Filter Bank | | Synthesis Filter Bank | |
|---|---|---|---|---|
| $n$ | $h(n)$ | $g(n-1)$ | $h'(n)$ | $g'(n+1)$ |
| 0 | 0.603 | 1.115 | 1.115 | 0.603 |
| $\pm 1$ | 0.267 | $-0.591$ | 0.591 | $-0.267$ |
| $\pm 2$ | $-0.078$ | $-0.058$ | -0.058 | $-0.078$ |
| $\pm 3$ | $-0.017$ | 0.091 | $-0.091$ | 0.017 |
| $\pm 4$ | 0.027 | | | 0.027 |

to wavelet synthesis from the upsampled DCT blocks. This method is referred to as the *wavelet to block transcoding* (WBDT) [148] technique. We highlight only the specific features of this computation under this context. For the sake of brevity and convenience, initially let us keep our discussion restricted to the 1-D with reference to the Daubechies 9/7 filter bank. The Corresponding impulse responses of analysis and synthesis filters of the 9/7 filter bank are shown in Table 6.1 (reproduced from Table 2.7 of Chapter 2). The computation in the DCT domain is carried out as discussed in Chapter 3. Consider $\mathbf{X}_{ae}^{(i)}$ and $\mathbf{X}_{do}^{(i)}$ are the upsampled DCT coefficients (of length $2N$) from the $i$th input blocks of analysis and detail coefficients $\mathbf{x}_a^{(i)}$ and $\mathbf{x}_d^{(i)}$, respectively (each of length $N$). The synthesis of these coefficients in the DCT domain takes place by exploiting the convolution-multiplication properties of the block DCT (refer to Chapter 3).

As lengths of right halves of both the symmetric filters are less than 8 (for 9/7 filter bank), we take care of the boundary effect of the convolution operation, by composing three adjacent upsampled DCT blocks into one larger type-II DCT block. The coefficients are then multiplied with the type-I DCT of the respective right half of the filter response. Consider $H'_I$ as the diagonal matrix whose elements are type-I $6N$-point DCT coefficients of right halves of $h'(n)$ (extended by padding with zeroes). Using the notation of the Chapter 3, we write it as

$$H'_I = \sqrt{12N}\mathbb{D}(\{\mathbf{C_{1e}}(\mathbf{h}'^+\}_0^{6N-1}).$$ (6.11)

Similarly, for the high-pass synthesis filter, we write it as

$$G'_I = \sqrt{12N}\mathbb{D}(\{\mathbf{C_{1e}}(\mathbf{g}'^+\}_0^{6N-1}).$$ (6.12)

In that case, the synthesized DCT coefficients are given by

$$\mathbf{Y_L^i} = PS_L \begin{bmatrix} \mathbf{x}_a^{(i-1)} \\ \mathbf{x}_a^{(i)} \\ \mathbf{x}_a^{(i+1)} \end{bmatrix},$$ (6.13)

and,

$$\mathbf{Y_H^i} = PS_H \begin{bmatrix} \mathbf{x_d^{(i-1)}} \\ \mathbf{x_d^{(i)}} \\ \mathbf{x_d^{(i+1)}} \end{bmatrix}. \tag{6.14}$$

and,

$$\mathbf{Y^i} = \mathbf{Y_L^i} + \mathbf{Y_H^i}, \tag{6.15}$$

where $P = \begin{bmatrix} 0_8 & 1_8 & 0_8 \end{bmatrix}$. $S_H$ and $S_L$ are given by the following expression:

$$S_L = A_{(3,2N)}^T H_I' A_{(3,2N)} \begin{bmatrix} B_{ue}^{(2N,N)} & 0_{2N \times N} & 0_{2N \times N} \\ 0_{2N \times N} & B_{ue}^{(2N,N)} & 0_{2N \times N} \\ 0_{2N \times N} & 0_{2N \times N} & B_{ue}^{(2N,N)} \end{bmatrix}, \tag{6.16}$$

$$S_H = A_{(3,2N)}^T G_I' A_{(3,2N)} \begin{bmatrix} B_{uo}^{(2N,N)} & 0_{2N \times N} & 0_2N \times N \\ 0_{2N \times N} & B_{uo}^{(2N,N)} & 0_{2N \times N} \\ 0_{2N \times N} & 0_{2N \times N} & B_{uo}^{(2N,N)} \end{bmatrix}, \tag{6.17}$$

where $0_{2N \times N}$ is the zero matrix of size $2N \times N$. The matrices $S_L$ and $S_H$ are of size $6N \times 3N$. Finally, as the central DCT block is to be retained as output, the cost of computation may be reduced by considering only the submatrix of $A^T$. Let $b_0^{(2N,6N)}$, $b_1^{(2N,6N)}$, and $b_2^{(2N,6N)}$ be submatrices of $A_{(3,2N)}^T$, each of size $2N \times 6N$ as shown in the following:

$$A_{(3,2N)}^T = \begin{bmatrix} b_0^{(2N,6N)} \\ b_1^{(2N,6N)} \\ b_2^{(2N,6N)} \end{bmatrix}. \tag{6.18}$$

The central output DCT block is computed using only the $b_1^{(2N,6N)}$ submatrix. Hence, the DCT domain filtering matrices, Eqs. (6.16) and (6.17), are modified as

$$S_L^{(2N,3N)} = b_1^{(2N,6N)} H_I' A_{(3,2N)} \begin{bmatrix} B_{ue}^{(2N,N)} & 0_{2N \times N} & 0_{2N \times N} \\ 0_{2N \times N} & B_{ue}^{(2N,N)} & 0_{2N \times N} \\ 0_{2N \times N} & 0_{2N \times N} & B_{ue}^{(2N,N)} \end{bmatrix}, \tag{6.19}$$

$$S_H^{(2N,3N)} = b_1^{(2N,6N)} G_I' A_{(3,2N)} \begin{bmatrix} B_{uo}^{(2N,N)} & 0_{2N \times N} & 0_2N \times N \\ 0_{2N \times N} & B_{uo}^{(2N,N)} & 0_{2N \times N} \\ 0_{2N \times N} & 0_{2N \times N} & B_{uo}^{(2N,N)} \end{bmatrix}, \tag{6.20}$$

where matrices $S_L^{(2N,3N)}$ and $S_H^{(2N,3N)}$ are of size $2N \times 3N$. For a given synthesis filter bank ($h'(n)$ and $g'(n)$), the filtering matrices $S_L^{(2N,3N)}$ and $S_H^{(2N,3N)}$

are usually precomputed. Typical examples of $S_L^{(8,12)}$ and $S_H^{(8,12)}$ using 9/7 filters are shown in Eqs. (6.21) and (6.22). [1]

$$
S_L^{(8,12)} =
\begin{bmatrix}
0 & 0 & 0 & -.0076 & .0795 & .1067 & .1021 & .1097 & .0226 & -.0047 & 0 & 0 \\
0 & 0 & 0 & -.0095 & .1108 & .0889 & -.0290 & -.1330 & -.0347 & .0065 & 0 & 0 \\
0 & 0 & 0 & -.0064 & .0984 & -.0514 & -.1387 & .0638 & .0404 & -.0061 & 0 & 0 \\
0 & 0 & 0 & -.0022 & .0663 & -.1232 & .0729 & .0234 & -.0428 & .0055 & 0 & 0 \\
0 & 0 & 0 & .0017 & .0250 & -.0581 & .0628 & -.0645 & .0378 & -.0047 & 0 & 0 \\
0 & 0 & 0 & .0042 & -.0034 & .0118 & -.0345 & .0449 & -.0265 & .0037 & 0 & 0 \\
0 & 0 & 0 & .0045 & -.0099 & .0125 & -.0041 & -.0145 & .0141 & -.0025 & 0 & 0 \\
0 & 0 & 0 & .0029 & -.0050 & .0020 & .0008 & .0033 & -.0052 & .0013 & 0 & 0
\end{bmatrix},
\tag{6.21}
$$

$$
S_H^{(8,12)} =
\begin{bmatrix}
0 & 0 & -.0022 & .0154 & -.0018 & .0014 & .0022 & -.0154 & .0018 & -.0014 & 0 \\
0 & 0 & -.0028 & .0225 & -.0042 & .0017 & -.0023 & .0234 & -.0035 & .0019 & 0 \\
0 & 0 & -.0019 & .0222 & -.0096 & .0117 & .0060 & -.0321 & .0055 & -.0018 & 0 \\
0 & 0 & -.0006 & .0172 & 0 & .0208 & -.0345 & .0459 & -.0068 & .0016 & 0 \\
0 & 0 & .0005 & .0091 & .0353 & -.0401 & .0410 & -.0506 & .0062 & -.0014 & 0 \\
0 & 0 & .0012 & .0033 & .0632 & -.0559 & .0146 & .0414 & -.0040 & .0011 & 0 \\
0 & 0 & .0013 & .0018 & .0619 & .0270 & -.0647 & -.0280 & .0015 & -.0007 & 0 \\
0 & 0 & .0008 & .0015 & .0388 & .0692 & .0592 & .0152 & -.0002 & .0004 & 0
\end{bmatrix}.
\tag{6.22}
$$

### 6.1.1.5  Transcoding in 2-D

In 2-D, as the FIR responses are separable, the above operations are performed in two stages, first by applying 1-D transcoding to the input wavelet block columns and then to rows of the resulting blocks. Let us consider the $LL$ subband and represent the $(i, j)th$ $N \times N$ block of wavelet coefficients in this band as $w_{LL}^{(i,j)}$ (and its matrix form as $\mathbf{w}_{LL}^{(i,j)}$). Hence, the computational steps for obtaining the $2N \times 2N$ synthesized DCT blocks in 2-D are shown below.

$$
\mathbf{Z}_{LL}^{(i,j)} = S_L^{(2N,3N)}
\begin{bmatrix}
\mathbf{w}_{(LL)}^{(i-1,j)} \\
\mathbf{w}_{(LL)}^{(i,j)} \\
\mathbf{w}_{(LL)}^{(i+1,j)}
\end{bmatrix},
\tag{6.23}
$$

$$
\mathbf{Y}_{LL}^{(i,j)} = \begin{bmatrix} \mathbf{Z}_{LL}^{(i,j-1)} & \mathbf{Z}_{LL}^{(i,j)} & \mathbf{Z}_{LL}^{(i,j+1)} \end{bmatrix} S_L^{(2N,3N)^T}.
\tag{6.24}
$$

where $\mathbf{Y}_{LL}^{(i,j)}$ is the transcoded synthesized block corresponding to the $LL$ band. In the same way, synthesis of other three subbands, namely, $LH$, $HL$, and $HH$, are carried out using the transcoding matrices $S_L^{(2N,3N)}$ and $S_H^{(2N,3N)}$. Finally, all these synthesized subband blocks are summed up to form the final synthesized output block. In the context of transcoding of a JPEG2000 image to JPEG, the value of $N$ is 4. It has been reported in [148] that the computational cost per pixel of the transcoded block of size $8 \times 8$ is $7.5M + 6.5A$, which provides around 7.61% savings in the computation compared to the spatial domain operations using the lifting-based wavelet synthesis.

---

[1] The elements in these matrices are to be scaled by $\sqrt{48}$.

The above scheme is extended also for transcoding multi-level wavelet decomposed images. There the upsampled DCT of the $LL$ band at any level is formed from its next higher level subbands as discussed previously. It is observed that this scheme is not efficient compared to the spatial domain operations using lifting-based wavelet synthesis. However, if we use *a hybrid of lifting based IDWT and DCT-based filtering* there is marginal reduction in the computation compared to the spatial domain approach [148].

## 6.1.2  DCT to DWT

For transcoding DCT to DWT, we need to carry out analysis of DCT blocks with a wavelet filter bank for obtaining respective wavelet coefficients in different subbands. As shown in Figure 6.1, the transcoding operations involve the following:

1. Filtering with DCT blocks using respective analysis filters ($h(n)$ and $g(n)$).

2. A filtered DCT block should be split into two halves, one with the *even downsampling* and the other with the *odd downsampling* (refer to Section 2.2.6.5 of Chapter 2).

In the next section, we first discuss how the splitting and downsampling could be performed on a DCT block to provide the result in the spatial domain.

### 6.1.2.1  Even and Odd Downsampling of DCT Blocks

The following two theorems state how the downsampling operation could be performed with a DCT block to provide the result in the spatial domain. For simplicity, we first consider the problem in 1-D. Later, the concepts are extended to 2-D.

**Theorem 6.3** *Let $x_{de}(m)$, $m = 0, 1, 2, \ldots, N/2 - 1$, (refer to Eq. (2.144) in Section 2.2.6.5 of Chapter 2) be the downsampled sequence formed by even indexed samples of $x(n)$, $n = 0, 1, \ldots, N - 1$ whose DCT is given by $X(k), k = 0, 1, \ldots, N - 1$. Then $x_{de}(m)$ could be obtained from $X(k)$ using the following expression:*

$$x_{de}(m) = \sqrt{\frac{2}{N}} \alpha(k) \sum_{k=0}^{N-1} X^{(N)}(k) \cos\left(\frac{(m + \frac{1}{4})\pi k}{\frac{N}{2}}\right), \ 0 \le m \le \frac{N}{2} - 1.$$

$$(6.25)$$

*where $\alpha(k)$ is defined in Eq. (2.89) in Chapter 2.*

**Proof:** For the sake of completeness the expression of IDCT of $X(k)$ is shown below.

$$C_{2e}^{-1}\{X(k)\} =$$

$$x(n) = \sqrt{\frac{2}{N}}\alpha(k)\sum_{k=0}^{N-1} X(k)\cos\left(\frac{(2n+1)\pi k}{2N}\right), 0 \le n \le N-1$$

$$(6.26)$$

where $\alpha(k)$ in Eq. (6.26) is $\sqrt{\frac{1}{2}}$ for $k = 0$, otherwise its value is 1.

Substituting $n$ with $2n$ in the above expression of IDCT, we obtain the following:

$$x(2n) = \sqrt{\frac{2}{N}}\alpha(k)\sum_{k=0}^{N-1} X(k)\cos\left(\frac{(2.2n+1)\pi k}{2N}\right), 0 \le 2n \le N-1.$$

From the above, renaming the variable $n$ as $m$, the following expression is obtained:

$$x_{de}(m) = \sqrt{\frac{2}{N}}\alpha(k)\sum_{k=0}^{N-1} X(k)\cos\left(\frac{\pi(m+\frac{1}{4})k}{\frac{N}{2}}\right), 0 \le m \le \frac{N}{2}-1.$$

□

**Theorem 6.4** Let $x_{do}(m)$, $m = 0, 1, 2, \ldots, N/2 - 1$, (refer to Eq. (2.145) in Section 2.2.6.5 of Chapter 2) be the downsampled sequence formed by odd indexed samples of $x(n)$, $n = 0, 1, \ldots, N-1$ whose DCT is given by $X(k), k = 0, 1, \ldots, N-1$. Then $x_{do}(m)$ could be obtained from $X(k)$ using the following expression:

$$x_{do}(m) = \sqrt{\frac{2}{N}}\alpha(k)\sum_{k=0}^{N-1} X(k)\cos\left(\frac{(m+\frac{3}{4})\pi k}{\frac{N}{2}}\right), 0 \le m \le \frac{N}{2}-1. \quad (6.27)$$

where $\alpha(k)$ is defined in Eq. (2.89) in Chapter 2.

**Proof:** In this case substituting $n$ with $2n+1$ in Eq. (6.26), we obtain the following expression:

$$x(2n+1) = \sqrt{\frac{2}{N}}\alpha(k)\sum_{k=0}^{N-1} X(k)\cos\left(\frac{(2(2n+1)+1)\pi k}{2N}\right),$$

$$0 \le (2n+1) \le N-1.$$

From the above, let us rename the term $n$ as $m$. Hence,

$$x_{do}(m) = \sqrt{\frac{2}{N}}\alpha(k)\sum_{k=0}^{N-1} X(k)\cos\left(\frac{(m+\frac{3}{4})\pi k}{N/2}\right), 0 \le m \le \frac{N}{2}-1.$$

The above theorems provide downsampled sequences, each of length $\frac{N}{2}$, from $N$-point type-II DCT block directly in the spatial domain. This reduces the computation for obtaining these sequences from a DCT block. Typically, for $N = 8$, conversion matrices for computing a downsampled block of length 4, are shown in Eqs. (6.28) and (6.29). They are denoted as $B_{de}^{(4,8)}$ and $B_{do}^{(4,8)}$, respectively.

$$B_{de}^{(4,8)} = \begin{bmatrix} 0.3536 & 0.4904 & 0.4619 & 0.4157 & 0.3536 & 0.2778 & 0.1913 & 0.0975 \\ 0.3536 & 0.2778 & -0.1913 & -0.4904 & -0.3536 & 0.0975 & 0.4619 & 0.4157 \\ 0.3536 & -0.0975 & -0.4619 & 0.2778 & 0.3536 & -0.4157 & -0.1913 & 0.4904 \\ 0.3536 & -0.4157 & 0.1913 & 0.0975 & -0.3536 & 0.4904 & -0.4619 & 0.2778 \end{bmatrix},$$

$$\tag{6.28}$$

$$B_{do}^{(4,8)} = \begin{bmatrix} 0.3536 & 0.4157 & 0.1913 & -0.0975 & -0.3536 & -0.4904 & -0.4619 & -0.2778 \\ 0.3536 & 0.0975 & -0.4619 & -0.2778 & 0.3536 & 0.4157 & -0.1913 & -0.4904 \\ 0.3536 & -0.2778 & -0.1913 & 0.4904 & -0.3536 & -0.0975 & 0.4619 & -0.4157 \\ 0.3536 & -0.4904 & 0.4619 & -0.4157 & 0.3536 & -0.2778 & 0.1913 & -0.0975 \end{bmatrix}.$$

$$\tag{6.29}$$

Hence, even and odd downsampled blocks from the DCT block $\mathbf{X}$ are obtained as follows:

$$\begin{aligned} \mathbf{x_{de}} &= B_{de}^{(4,8)} \mathbf{X}, \\ \mathbf{x_{oe}} &= B_{oe}^{(4,8)} \mathbf{X}. \end{aligned} \tag{6.30}$$

In the above equation, $\mathbf{x_{de}}$ and $\mathbf{x_{do}}$ are of length 4, while $\mathbf{X}$ is of length 8.

### 6.1.2.2　Even and Odd Downsampling in 2-D

The extension of the above concepts to 2-D is trivial. Consider $X$ be an $8 \times 8$ DCT block. Its downsampled block in the $LL$ subband, $\mathbf{x_{LL}}$ of size $4 \times 4$ is computed as

$$\mathbf{x_{LL}} = B_{de}^{(4,8)} X B_{de}^{(4,8)^T}. \tag{6.31}$$

Similarly, $\mathbf{x_{HH}}$ for the $HH$-subband is computed as

$$\mathbf{x_{HH}} = B_{do}^{(4,8)} X B_{do}^{(4,8)^T}. \tag{6.32}$$

Similarly, the other two subbands ($LH$ and $HL$) are also computed.

### 6.1.2.3　Wavelet Analysis in the DCT Domain

For obtaining wavelet coefficients from DCT blocks, analysis filters may be applied directly in the DCT domain. While filtering DCT blocks at the boundaries, neighboring blocks outside the image are formed through symmetric extension using the flipping transformation discussed in Section 3.4.1 of Chapter 3. Otherwise, the basic filtering computation in the block DCT space remains the same. Let us briefly discuss the computation involved in this process.

The diagonal matrices formed by the type-I DCTs of analysis filters, $h(n)$ and $g(n)$, are given below.

$$H_I = \sqrt{6N} \mathbb{D}(\{\mathbf{C_{1e}}(\mathbf{h^+}\}_0^{3N-1}), \tag{6.33}$$

$$G_I = \sqrt{6N}\mathbb{D}(\{\mathbf{C_{1e}}(\mathbf{g}^+\}_0^{3N-1}).\tag{6.34}$$

Let the $i$th $N$-point DCT block be denoted as $\mathbf{X}^{(i)}$. The composite transcoding operations from DCT blocks are given below.

$$
\begin{aligned}
\mathbf{x_a}^{(i)} &= R_L^{(N/2,3N)} \begin{bmatrix} \mathbf{X}^{(i-1)} \\ \mathbf{X}^{(i)} \\ \mathbf{X}^{(i+1)} \end{bmatrix}, \\
\mathbf{x_d}^{(i)} &= R_H^{(N/2,3N)} \begin{bmatrix} \mathbf{X}^{(i-1)} \\ \mathbf{X}^{(i)} \\ \mathbf{X}^{(i+1)} \end{bmatrix},
\end{aligned}
\tag{6.35}
$$

where $\mathbf{x_a}^{(i)}$ and $\mathbf{x_d}^{(i)}$ are wavelet approximation and detailed coefficients, each of size $\frac{N}{2}$. $R_L^{(N/2,3N)}$, and $R_H^{(N/2,3N)}$ are as follows:

$$R_L^{(N/2,3N)} = B_{de}^{(N/2,N)} b_1^{(N,3N)} H_I A_{(3,N)},\tag{6.36}$$

$$R_H^{(N/2,3N)} = B_{do}^{(N/2,N)} b_1^{(N,3N)} G_I A_{(3,N)}.\tag{6.37}$$

As before, $b_1^{(N,3N)}$ represents the submatrix of $A_{(3,2N)}^{-1}$ (refer to Eq. (6.18)). Above computation is extended in 2-D following the similar technique discussed in the previous section on transcoding DWT to DCT coefficients. The computational cost as reported in [147] is found to be marginally more efficient than the transcoding through the spatial domain using lifting-based schemes. The composite costs for both these operations are given as 29.125 and 31.39 per pixel of the image.

### 6.1.3 DCT to ICT

The integer cosine transform (ICT) matrix used in H.264 video coding is of size $4 \times 4$ and is denoted by $T_4$ in this text (refer to Section 2.2.4.9 of Chapter 2). For the purpose of transcoding, we need to convert a DCT block (in 1-D) to *two* blocks of ICT. The conversion matrix $C_T$ is given by

$$C_T = \begin{bmatrix} T_4 & 0_4 \\ 0_4 & T_4 \end{bmatrix} C_8^T.\tag{6.38}$$

In the above equation, $C_8$ is the 8-point DCT matrix in the above equation, and $0_8$ is the $8 \times 8$ zero matrix. The elements of $C_T$ follow a symmetry, and a few of them are zeroes as shown below.

$$C_T = \begin{bmatrix} 1.414 & 1.281 & 0 & -0.450 & 0 & 0.301 & 0 & -0.255 \\ 0 & 0.924 & 2.230 & 1.780 & 0 & -0.864 & -0.159 & 0.482 \\ 0 & -0.106 & 0 & 0.726 & 1.414 & 1.086 & 0 & -0.531 \\ 0 & 0.117 & 0.159 & -0.092 & 0 & 1.038 & 2.230 & 1.975 \\ 1.414 & -1.281 & 0 & 0.450 & 0 & -0.301 & 0 & 0.255 \\ 0 & 0.924 & -2.230 & 1.780 & 0 & -0.864 & 0.159 & 0.482 \\ 0 & 0.106 & 0 & -0.726 & 1.414 & -1.086 & 0 & 0.531 \\ 0 & 0.117 & -0.159 & -0.092 & 0 & 1.038 & -2.230 & 1.975 \end{bmatrix}.$$

$$(6.39)$$

By exploiting symmetry in the occurrences of the elements, we require only 22 multiplications and 22 additions for the transcoding in 1-D. In 2-D, a DCT block $\mathbf{X}$ is transformed as $C_T \mathbf{X} C_T^T$ with 352 multiplications and 352 additions.

### 6.1.4   ICT to DCT

In the same way, we obtain the transcoding matrix for converting *two* blocks of ICT in 1-D into *one* DCT block. The corresponding conversion matrix $T_C$ is given below.

$$T_C = C_8 \begin{bmatrix} T_4^{-1} & 0_4 \\ 0_4 & T_4^{-1} \end{bmatrix}. \tag{6.40}$$

Magnitudes of the elements of $T_C$ also show center-column-based symmetry.

$$T_C = \begin{bmatrix} 0.354 & 0 & 0 & 0 & 0.354 & 0 & 0 & 0 \\ 0.320 & 0.092 & -0.026 & 0.012 & -0.320 & 0.092 & 0.026 & 0.012 \\ 0 & 0.223 & 0 & 0.016 & 0 & -0.223 & 0 & -0.016 \\ -0.112 & 0.178 & 0.181 & -0.009 & 0.112 & 0.178 & -0.181 & -0.009 \\ 0 & 0 & 0.354 & 0 & 0 & 0 & 0.354 & 0 \\ 0.075 & -0.086 & 0.272 & 0.104 & -0.075 & -0.086 & -0.272 & 0.104 \\ 0 & -0.016 & 0 & 0.223 & 0 & 0.016 & 0 & -0.223 \\ -0.064 & 0.048 & -0.133 & 0.197 & 0.064 & 0.048 & 0.133 & 0.197 \end{bmatrix}$$

$$(6.41)$$

In this case it is required to perform 22 multiplications and 34 additions for the conversion in 1-D. In 2-D, these numbers are 352 and 504, respectively.

## 6.2   Image Transcoding: JPEG2000 to JPEG

In this section, we consider a typical example of transcoding of images, namely, transcoding of a JPEG2000-encoded image into the DCT-based JPEG compression format as reported in [147]. The transcoding tasks involve the conversion of DWT coefficients into $8 \times 8$ blocks of DCT coefficients. However, the process also requires entropy decoding and dequantization to obtain the DWT coefficients. Further, DCT coefficients are to be quantized and encoded according to the JPEG standard, as discussed in Chapter 1. In the study reported in [147], simplified implementations of JPEG and JPEG2000 coding systems as discussed in [53] were used. Using these implementations, they achieve compression ratios differing approximately by a factor of two compared to standard encoders. However, they keep the relative performance of these two schemes at par with the exact implementations of the compression schemes. Moreover, an estimate of the true compression ratio is also obtained by multiplying it by a factor of 2. In the study, images are decomposed into three levels using the 9/7 analysis filters. It achieves a compression of 0.943 bpp (for the image *Lena*) with the quantization steps of ten subbands as given below.

$$\triangle_b = [16.06, 16.06, 32.13, 8.03, 8.03, 16.06, 4.02, 4.02, 8.03, 2.01]. \qquad (6.42)$$

During decoding, the DWT coefficients of these subbands are dequantized and subjected to the multilevel transcoding of wavelet coefficients into the $8 \times 8$ block DCT space. Finally, the resulting $8 \times 8$ DCT blocks are quantized and encoded using standard Huffman coding to obtain the transcoded image in JPEG format. The study considered also a few approaches of transcoding, including the DCT filtering based approach discussed in Section 6.1.1.4 (referred to as *wavelet to block DCT transcoding* (WBDT)). A brief overview of different techniques is presented below.

### 6.2.1 Transcoding with WBDT

The WBDT technique is discussed in Section 6.1.1.4. We also consider two variations of these approaches. If all the DWT subbands are transcoded into DCT blocks, the approach is referred to as WBDT-all. On the other hand, if only the low-frequency subband (LL) is used, it is called WBDT-LL.

### 6.2.2 Transcoding with Wavelet Doubling

In this technique, only the LL band is synthesized, and then it is transformed into the block DCT space. This transcoding technique requires only LL subband reconstruction and is referred to as J2K-LL.

### 6.2.3 Transcoding with DCT Domain Doubling

In this approach, an efficient DCT domain doubling algorithm [36] is used for doubling the *LL* band, First, the LL band is transformed into $4 \times 4$ block DCT space. Next, each of these blocks is converted into an $8 \times 8$ block using subband approximation. This transcoding technique is called *doubling algorithm for transcoding* (DAT).

Apart from the above, we also refer to the spatial domain transcoding

Table 6.2: Per-pixel computational cost of different transcoding approaches.

| No. of subbands | Method | No. of operations | | CC |
|---|---|---|---|---|
| | | $\mathcal{M}$ | $\mathcal{A}$ | |
| One | J2K-LL (FB) | 8.375 | 13.28 | 38.40 |
| (LL) | J2K-LL (LS) | **5.875** | 11.78 | 29.40 |
| | WBDT-LL | 7.00 | **6.00** | **27.00** |
| | DAT | **0.25** | 1.15 | **1.90** |
| Four | ST-FB | 25.12 | 39.84 | 115.20 |
| (LL, HL, LH, HH) | ST-LS | **17.62** | 35.34 | 88.21 |
| | WBDT-all | 21.00 | **18.00** | **81.00** |

*Source:* From [147] with permission from the author.

technique as ST, which has two implementations. The one using filter banks is referred to as ST-FB. The other implementation is by the lifting scheme (referred to as ST-LS). The computational costs for different techniques as reported in [147] are shown in Table 6.2.

We observe that the DAT algorithm is computationally most efficient among the transcoding techniques with single subband considered in the study. The WBDT-LL transcoding and WBDT-all are also computationally efficient compared to the corresponding spatial domain techniques (ST). It is of interest to know the performance in terms of quality of transcoded images from these different approaches. Let us consider a few performance metrics as proposed in [147] for this purpose.

### 6.2.4    Performance Metrics for Transcoding Schemes

In [147] three performance metrics are defined with respect to the rate distortion curve obtained from a transcoding scheme. In Figure 6.2(a), a typical rate distortion curve for transcoding a JPEG2000 image (*Lena* compressed at 0.55 bpp) into JPEG encoding using the WBDT algorithm is shown. In this study, the JPEG2000 decoded image acts as the reference image. With respect to this curve, the transcoding measures are defined below.

1. **Guaranteed PSNR** ($M_G$): It is the minimum PSNR value with respect to the JPEG2000 decoded image to be achieved by the transcoder. This is specified per the user's requirement.

2. **Equivalent rate** ($\rho_{eq}$): It is the minimum compression rate of the transcoded JPEG image providing at least the specified guaranteed PSNR with respect to the JPEG2000 decoded image.

3. **Equivalent PSNR** ($M_{eq}$): It is the PSNR with respect to the original image at the equivalent rate (given a guaranteed PSNR level).

The above measures are illustrated with the help of plots as shown in Figure 6.2 (**see color insert**). According to the plot in Figure 6.2(a), given the guaranteed PSNR, $M_G$ as 40 dB, the equivalent rate is obtained as 1.25 bpp in the transcoder. The equivalent rates for the transcoded JPEG image at different JPEG2000 compression levels are shown in Figure 6.2(b). The performance curve varies with different values of guaranteed PSNR. For $M_G$ at guaranteed 40 dB, the variation of equivalent PSNR and equivalent rates is shown in Figure 6.2(c). At the PSNR of 32 dB, the transcoder performance becomes almost the same as the JPEG2000 performance. For other schemes, we also observe their performances against these metrics. For example, for the transcoding scheme DAT, the equivalent rate becomes almost the same as the original compression rate (refer to Figure 6.2(b)) at a guaranteed PSNR of 30 dB, which is kept at a lower value as the reconstruction is possible only from the LL band. It is observed that performance curves of the spatial domain transcoding ( ST) and the transform domain technique (WBDT-all) are almost identical, though the latter runs much faster than the former.

Another useful feature of these schemes is to observe the maximum PSNR value that could be achieved from the transcoded image with respect to the original image. This value is less than the original PSNR value of the encoded image (with JPEG2000 in this case). In Figure 6.3 (**see color insert**), maximum PSNR values obtained at varying level of compression are shown for different images. As typical examples, transcoded images of *Lena* and their maximum PSNR values obtained by different transcoding schemes are shown in Figure 6.4. It is reported that at higher bpp, the maximum PSNR values obtained by the DAT are 3–4 dB less compared to the WBDT-LL and 4–6 dB less compared to the *spatial* transcoding (ST).

## 6.3   Video Downscaling

In different applications such as video browsing, picture in picture, video conferencing, etc., the video is required to be transcoded at lower bit rates and reduced frame size. This task is known as *video downscaling*. The brute force approach to performing this operation is to decode each frame of the input video, downscale each of them spatially and, finally, re-encode the whole downscaled raw video at a lower bit rate. This technique is the spatial domain downscaling. However, it is very time consuming and computationally inefficient to meet the requirements of real-time applications. In addition to the DCT/IDCT operations, motion estimation during re-encoding of downscaled video puts a heavy burden in the computation. In this section, we discuss how compressed domain operations could greatly reduce the computational cost to this end. In our discussion, we mostly restrict ourselves to the downscaling of an MPEG video stream. From an MPEG video, using Huffman decoding

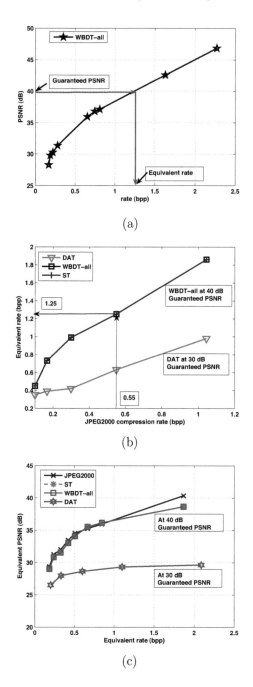

(a)

(b)

(c)

**Figure 6.2**: JPEG2000 to JPEG transcoding for Lena image using the WBDT technique: (a) rate distortion for transcoded image with respect to the JPEG2000 decoded image (for 0.55 bpp encoded Lena image), (b)equivalent rate of the transcoder versus JPEG2000 compression rate, and (c)equivalent PSNR versus equivalent rate. (From [147] with permission from the author.)

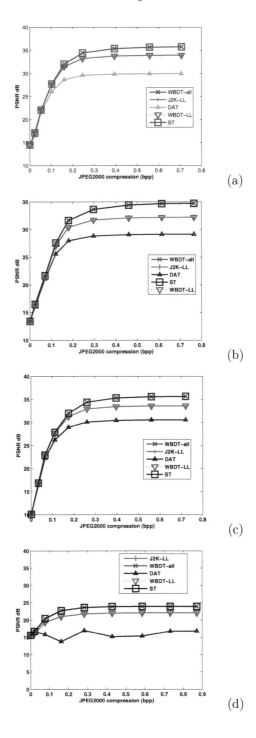

**Figure 6.3**: Maximum PSNR values versus compression ratio:(a) Lena, (b) Peppers, (c) Girl, and (d) Baboon. (With permission from [147].)

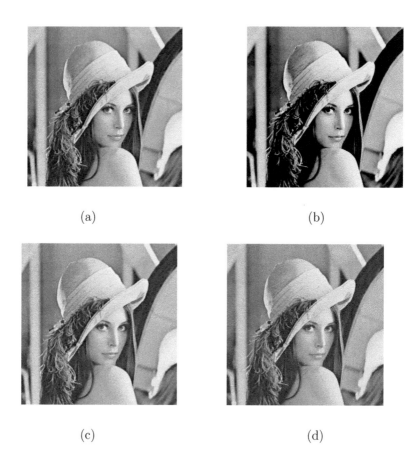

(a)                                        (b)

(c)                                        (d)

**Figure 6.4**: Results for the JPEG2000 encoded Lena image:(a) JPEG2000 encoded image at bpp = 0.7062, PSNR = 38.47 dB, (b) JPEG encoded by WBDT-all maximum PSNR = 35.21 dB (at bpp = 0.9 or higher), (c) JPEG encoded image by WBDT-LL maximum PSNR = 33.94 dB (at bpp = 0.666 or higher), and (d) JPEG encoded image by DAT maximum PSNR = 29.95 dB (at bpp = 0.784 or higher). (With permission from [147].)

and dequantization of DCT coefficients, we finally obtain a set of $8 \times 8$ DCT blocks for each of its frames. These DCT blocks are operated in the process of downscaling of video frames. It is known that an MPEG video has two types of frames, namely, the intraframe and interframe. Intraframes are the same as, those specified in the JPEG compression standard. Hence, all the techniques of image downsizing are suitable for the purpose of downsampling. However the inter frames are motion-compensated for obtaining a higher degree of compression. Hence, we have to reconstruct these motion-compensated frames using *inverse motion compensation* (IMC) techniques before downscaling. Subsequently, for efficient encoding of the sequence of downscaled frames with motion compensation, we use the motion vectors for each macroblock of the interframe. This task is known as *refinement of motion vectors*. In the following subsections, we first discuss how these two tasks are performed efficiently in the block DCT space. Next, we discuss the overall video downscaling technique in the block DCT space.

### 6.3.1 Inverse Motion Compensation

Inverse motion compensation is the process by which a predictive frame in a video stream is converted into an intra (or $I$) frame. In our discussion, we restrict ourselves to forward predictive frames only, which are also known as $P$ frames. In the MPEG standard, motion estimation and compensation are carried out over a group of four adjacent $8 \times 8$ DCT blocks, which is called a macroblock. In this computation, a prediction $M'$ of each macroblock $M$ in the current frame is made from the previous encoded frame. It is the prediction error $E = M - M'$, which is encoded in the video stream. Though the prediction is carried out for a $16 \times 16$ macroblock, errors are encoded by applying $8 \times 8$ DCT to each of its component blocks. The computational challenge in the block DCT space, is to extract the DCT of the corresponding component block of $M'$ in the previous frame, so that it could be added to the DCT encoded error, resulting in recovery of transform coefficients of sample values. In that case, the block turns into an *intrablock*. This is the principle of inverse motion compensation (IMC). In the following, we review two approaches of the IMC. In one [99], a single $8 \times 8$ block is subjected to the IMC, whereas in the other approach [117], this is carried out for the whole macroblock at a time.

#### 6.3.1.1 Single Blockwise Inverse Motion Compensation

In this technique [20, 99] the IMC is performed on an $8 \times 8$ DCT block. As the computation is carried out for a single $8 \times 8$ block of the $P$ frame, we refer to this technique as *single blockwise motion compensation* (SBIMC). Let us assume that the intrablock coding is available for the previous frame, which acts as the reference for recovery of the true sample values. Given the motion

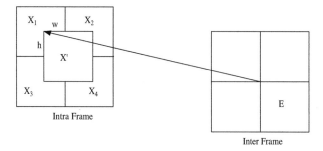

Figure 6.5: Single blockwise inverse motion compensation (SBIMC).

vector of this block, a possible configuration of its predicted $8 \times 8$ block with respect to the intrablocks in the previous frame is shown in Figure 6.5.

In general, the predicted block $X'$ may not align itself with any one of the $8 \times 8$ intra DCT blocks (that is, $X_1$, $X_2$, $X_3$, and $X_4$) in the reference frame as it may intersect with four $8 \times 8$ DCT blocks as shown in Figure 6.5. In the figure the $h$ and $w$ are vertical and horizontal components of the motion vector, respectively. Let $\mathbf{x_1}$, $\mathbf{x_2}$, $\mathbf{x_3}$, and $\mathbf{x_4}$ denote the corresponding blocks in the spatial domain. In that case, the overlapped $8 \times 8$ block $\mathbf{x'}$ corresponding to the motion vector is formed by Eq. (6.43) [99].

$$
\begin{aligned}
\mathbf{x'} &= \sum_{i=1}^{4} c_{i1}\mathbf{x_i}c_{i2} \\
&= U_h\mathbf{x_1}L_w + U_h\mathbf{x_2}U_{8-w} + L_{8-h}\mathbf{x_3}L_w + L_{8-h}\mathbf{x_4}U_{8-w}.
\end{aligned}
\tag{6.43}
$$

In the above equations, $c_{ij}$'s are defined by matrices $U_n$ or $L_n$. $U_n$ and $L_n$ for $n = 1, 2, ...8$ are defined below.

$$
\begin{aligned}
U_n &= \begin{bmatrix} 0_{n \times (8-n)} & I_n \\ 0_{8-n} & 0_{(8-n) \times n} \end{bmatrix}, \\
L_n &= \begin{bmatrix} 0_{(8-n) \times n} & 0_{8-n} \\ I_n & 0_{n \times (8-n)} \end{bmatrix}.
\end{aligned}
\tag{6.44}
$$

The DCT of the predicted block $\mathbf{x'}$ is obtained from Eq. (6.43) in the following form:

$$
\mathbf{X'} = C_8 \left( \sum_{i=1}^{4} c_{i1}C_8^T C_8 \mathbf{x_i} C_8^T C_8 c_{i2} \right) C_8^T.
\tag{6.45}
$$

We use the factorization of the $8 \times 8$ DCT matrix ($C_8$) for efficiently computing the above. From Section 2.2.4.8 of Chapter 2, the factorization of $C_8$ is given in the following form:

$$C_8 = DPB_1B_2MA_1A_2A_3. \tag{6.46}$$

Definitions of different factoring matrices are also given in that section. Hence, Eq. (6.45) is rewritten as

$$\mathbf{X}' = C_8[J_h B_2^T B_1^T P^T D(\mathbf{X_1} DPB_1 B_2 J_w^T + \mathbf{X_2} DPB_1 B_2 K_{8-w}^T)$$
$$+ K_{8-h} B_2^T B_1^T P^T D(\mathbf{X_3} DPB_1 B_2 J_w^T + \mathbf{X_4} DPB_1 B_2 K_{8-h}^T)]C_8^T, \tag{6.47}$$

where
$J_i = U_i(MA_1A_2A_3)^T$ , $i = 1, 2, 3, ..8$, and,
$K_i = L_i(MA_1A_2A_3)^T$, $i = 1, 2, 3, ..8$.
The above computations are carried out very efficiently by adopting the following:

1. Multiplications by elements of the diagonal matrix $D$ are absorbed in the MPEG quantization steps. Hence these multiplications are ignored.

2. As $P$ and $P^T$ are permutation matrices, there is no requirement for actual multiplication and addition operations. Multiplication with these matrices causes only changes in the order of the elements of the multiplicant.

In this computation, the total number of operations gets maximized for $h = w = 4$ as reported in [99], and it is found to be $320M + 1760A$. This implies that the computational cost of SBIMC is at most $5M + 27.5A$ operations per pixel of the input video frame.

### 6.3.1.2  Macroblockwise Inverse Motion Compensation

In [117] a modification of the above scheme is presented, which considers performing the computation over a macroblock. In an MPEG video stream, a macroblock contains four $8 \times 8$ blocks and motion vectors, and prediction errors are estimated for a $16 \times 16$ macroblock. The technique exploits this fact to reduce the overall computation for the *four blocks*.

In Figure 6.6, it is shown how the predicted $16 \times 16$ block (also called macroblock due to its size of $16 \times 16$) overlaps with the DCT blocks in the reference frame. Let $M'$ be the predicted macroblock in the reference frame, at a location $(r, c)$ as shown in the figure. Under this scenario, $M'$ is computed from the nine $8 \times 8$ DCT blocks. Let $x_i, 1 \leq i \leq 9$, be the adjacent blocks in the spatial domain as shown in Figure 6.6. In this case, we need to extract a $16 \times 16$ block from them. The relationship between the 9 DCT blocks and $M'$ is expressed in the following form:

$$\mathbf{m}' = c_r \begin{bmatrix} \mathbf{x_1} & \mathbf{x_2} & \mathbf{x_3} \\ \mathbf{x_4} & \mathbf{x_5} & \mathbf{x_6} \\ \mathbf{x_7} & \mathbf{x_8} & \mathbf{x_9} \end{bmatrix} c_c^T, \tag{6.48}$$

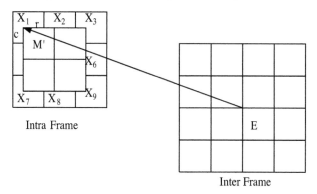

Figure 6.6: Macroblockwise inverse motion compensation (MBIMC).

where $\mathbf{m}'$ is the predicted macroblock in spatial domain and $c_i$ is a $16 \times 24$ matrix as defined below.

$$c_i = \begin{bmatrix} 0_{16 \times i} & I_{16} & 0_{16 \times (8-i)} \end{bmatrix}. \tag{6.49}$$

Note that there are 8 different $c_i$'s as $i$ varies from 0 to 7.

In the DCT domain, the above relationship (Eq. (6.48)) is expressed as

$$
\mathbf{M}' = \left( \begin{array}{cc} C_8 & 0_8 \\ 0_8 & C_8 \end{array} \right) \left\{ c_r \begin{bmatrix} C_8^T & 0_8 & 0_8 \\ 0_8 & C_8^T & 0_8 \\ 0_8 & 0_8 & C_8^T \end{bmatrix} \begin{bmatrix} \mathbf{X_1} & \mathbf{X_2} & \mathbf{X_3} \\ \mathbf{X_4} & \mathbf{X_5} & \mathbf{X_6} \\ \mathbf{X_7} & \mathbf{X_8} & \mathbf{X_9} \end{bmatrix} \begin{bmatrix} C_8 & 0_8 & 0_8 \\ 0_8 & C_8 & 0_8 \\ 0_8 & 0_8 & C_8 \end{bmatrix} c_c \right\} \left( \begin{array}{cc} C_8^T & 0_8 \\ 0_8 & C_8^T \end{array} \right).
\tag{6.50}
$$

By using the factorization of DCT in a similar manner (as is done for the SBIMC technique), it has been shown [116] that the MBIMC requires at the most $3.43M + 20.5A$ operations per pixel of the frame. Assuming cost of three additions are equivalent to one multiplication there is 27% reduction of the computational cost with respect to the SBIMC scheme.

### 6.3.1.3 Video Downscaling and IMC: Integrated Scheme

In [117] the MBIMC technique is integrated with the downsampling operation to provide the downsampled inverse motion compensated block as a result of this composite operation. Let the downsampling matrix operator be denoted as $D_{8 \times 16}$ following the same notation used in Section 5.1.1 of Chapter 5. The matrix $D_{8 \times 16}$ is given below.

$$
D_{8 \times 16} =
\begin{bmatrix}
0.5 & 0.5 & 0 & 0 & 0 & 0 & 0 & 0 & 0 & 0 & 0 & 0 & 0 & 0 & 0 & 0 \\
0 & 0 & 0.5 & 0.5 & 0 & 0 & 0 & 0 & 0 & 0 & 0 & 0 & 0 & 0 & 0 & 0 \\
0 & 0 & 0 & 0 & 0.5 & 0.5 & 0 & 0 & 0 & 0 & 0 & 0 & 0 & 0 & 0 & 0 \\
0 & 0 & 0 & 0 & 0 & 0 & 0.5 & 0.5 & 0 & 0 & 0 & 0 & 0 & 0 & 0 & 0 \\
0 & 0 & 0 & 0 & 0 & 0 & 0 & 0 & 0.5 & 0.5 & 0 & 0 & 0 & 0 & 0 & 0 \\
0 & 0 & 0 & 0 & 0 & 0 & 0 & 0 & 0 & 0 & 0.5 & 0.5 & 0 & 0 & 0 & 0 \\
0 & 0 & 0 & 0 & 0 & 0 & 0 & 0 & 0 & 0 & 0 & 0 & 0.5 & 0.5 & 0 & 0 \\
0 & 0 & 0 & 0 & 0 & 0 & 0 & 0 & 0 & 0 & 0 & 0 & 0 & 0 & 0.5 & 0.5
\end{bmatrix}
\tag{6.51}
$$

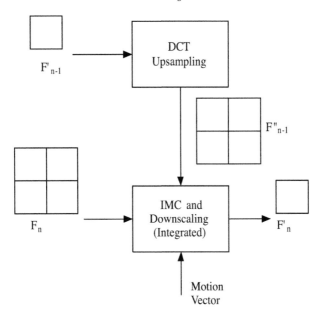

**Figure 6.7**: An integrated scheme for downscaling and inverse motion compensation.

From Eq. (6.50) we get the following expression for computing the downsampled block:

$$
\mathbf{M'_d} = \\
C_8 \left\{ D_{8 \times 16} c_r \begin{bmatrix} C_8^T & 0_8 & 0_8 \\ 0_8 & C_8^T & 0_8 \\ 0_8 & 0_8 & C_8^T \end{bmatrix} \begin{bmatrix} \mathbf{x_1} & \mathbf{x_2} & \mathbf{x_3} \\ \mathbf{x_4} & \mathbf{x_5} & \mathbf{x_6} \\ \mathbf{x_7} & \mathbf{x_8} & \mathbf{x_9} \end{bmatrix} \begin{bmatrix} C_8 & 0_8 & 0_8 \\ 0_8 & C_8 & 0_8 \\ 0_8 & 0_8 & C_8 \end{bmatrix} c_c D_{8 \times 16}^T \right\} C_8^T .
$$

$$(6.52)$$

As the IMC operation provides the downsampled intraframe in the block DCT space, for performing IMC of the next frame, we need to upsample it as shown in Figure 6.7. In this figure, $F'_{n-1}$ is the $(n-1)$th downscaled reference frame, and $F_n$ is the next interframe of the input video. For performing this operation, first the downsampled reference frame $(F'_{n-1})$ is upsampled by a factor of two to form the $(n-1)$th upsampled reference $F''_{n-1}$. Next, $F''_{n-1}$ is used as a reference to the integrated computation providing the donsampled frame $F'_n$. For converting an interframe immediately after an intraframe (I), this upsampling operation is not necessary. In the previous chapter, various such upsampling strategies were discussed including the efficient one [36] requiring $1.25M + 1.25A$ per pixel of the upsampled image. In [117], it is shown that this computation could be efficiently carried out with a computational cost of at the most $3.34M + 13.65A$ per pixel of the interframe.

## 6.3.2  Motion Vector Refinement

Motion vectors are computed for each macroblock as offsets of the predicted blocks with respect to their locations in the current frame. Using these motion vectors, prediction errors are obtained at the encoder, and motion compensation is carried out in the decoder. For higher compression in the downsampled video, we should carry out a similar computation. As recomputation of motion vectors is quite expensive, we may predict a motion vector of the downscaled video using the motion vectors of the original compressed video stream. This task is referred here as the *motion vector refinement*. *Four* adjacent macroblocks in the full-scale video are mapped into *one* macroblock in its downscaled version. Hence, in various approaches [57, 66, 129, 162], the refined motion vector in the downscaled block is computed from those four vectors of adjacent macroblocks of the original video (compressed with the MPEG video compression standard). Some of these approaches are discussed in the following subsections.

### 6.3.2.1  Adaptive Motion Vector Resampling (AMVR)

In the AMVR technique [129] the motion vector for the downsampled block $\overrightarrow{mv_d}$ is computed as a weighted average of the four motion vectors of the corresponding macroblocks of the original video. Let us denote the motion vectors of those macroblocks as $\overrightarrow{mv_i}$, $1 \leq i \leq 4$. Then $\overrightarrow{mv_d}$ is given by the following relationship:

$$\overrightarrow{mv_d} = \frac{1}{2} \frac{\sum_{i=1}^{4} \overrightarrow{mv_i} W_i}{\sum_{i=1}^{4} W_i}, \tag{6.53}$$

where $W_i$ is the weight of the motion vector for the $i$th vector.

In the above computation, weights are adaptively determined from the residual errors of the respective blocks. Ideally, the refined motion vector should be one, which minimizes (at least locally) the residual error of prediction in the downsampled sequence. The weights are chosen in such a way that the performance of the resulting vector is close to it. In [129], several heuristics have been explored to this end. For example, when all the weights are equal, the AMVR computation turns out to be a simple case of averaging, which is referred to as *align-to-average weighting* (AAW). However, it is found that in most cases there are better approximations than this. Another policy is choosing the refined motion vector from one of the $\overrightarrow{mv_i}$'s. For example, we choose the $i$th block, which has *maximum* or *minimum* residual error among these four. These strategies are known as *align-to-the-worst weighting* (AWW) or *align-to-the-best weighting* (ABW), respectively. Experimentally, it is observed that the AWW approach gives the best performance (reportedly on an average of a 0.5 dB improvement from others) in most cases out of the above three. However, there are other ways to determine the weights from the residual errors. For example, they may be made proportional to the number of nonzero quantized AC coefficients.

#### 6.3.2.2 Median Method

In this approach [127], a motion vector corresponding to one of *four* mac-roblocks in the full-scale video is chosen for the refinement operation. The criterion for the selection is that it should be the median vector among them. A median of a set of vectors is defined in the following. In this definition, let us restrict ourselves to the set of four vectors. However, the same concept is applicable to any number of vectors in the set.

Let $V = \{\vec{v_1}, \vec{v_2}, \vec{v_3}, \vec{v_4}\}$ form the set of motion vectors for adjacent blocks in the full-scale video. The distance between a vector $\vec{v_i} \in V$ and $V$ is defined as follows:

$$d(\vec{v_i}, V) = \sum_{j=1(j \neq i)}^{4} \parallel \vec{v_i} - \vec{v_j} \parallel, \qquad (6.54)$$

where $\parallel \vec{u} \parallel$ is the Euclidean norm of the vector $\vec{u}$.

The median vector $\vec{v_m}$ is defined as one of these vectors that has the least distance from $V$, that is,

$$\vec{v_m} = \arg \min_{\vec{v_i} \in V} \{d(\vec{v_i}, V)\}.$$

Finally, the motion vector $\vec{mv_d}$ in the downsampled frame is computed as $\frac{\vec{v_m}}{2}$.

#### 6.3.2.3 Nonlinear Motion Vector Resampling (NLMR) Method

In this approach [66], a likelihood measure for each candidate motion vector to be the representative of the set is estimated, and then the vector with the best likelihood score is chosen for the refinement operation. To perform this task, initially clusters of aligned motion vectors are computed. A pair of motion vectors belong to the same cluster if the cityblock [52] distance $d$ between them is less than 1. Let those vectors be represented as $(v_{x1}, v_{y1})$ and $(v_{x2}, v_{y2})$. The cityblock distance $d$ between them is computed as

$$d = \mid v_{x1} - v_{x2} \mid + \mid v_{y1} - v_{y2} \mid . \qquad (6.55)$$

In this method, finally each motion vector is associated with a set of four features, namely, the activity of macroblock $(A)$, quantization step size $(Q)$, size of the cluster to which it belongs $(C)$, and the magnitude of motion vector $(M)$. A likelihood score $L_i$ (initial value being set at zero) is computed for each $i$th motion vector according to the following rule:

*The $L_i$ is incremented whenever the i-th parameter $A, Q, C$ is highest or the i-th parameter $M$ is the lowest among the macroblocks.*

The motion vector corresponding to the highest likelihood score is chosen for the downscaled macroblock. However, if there is more than one motion vector with the same likelihood score, their weighted average provides the representative vector. It has been reported in [116] that the performance of all the three techniques mentioned before are more or less the same. However, the NLMR technique provides marginal improvement, when it is used for fast-moving videos.

### 6.3.3 Macroblock Type Declaration

Deciding the type of the macroblock in the downscaled video is also a task to be accomplished in the process of downsizing. In [56], a simple approach to selecting the type of a macroblock is presented. In this scheme, a downscaled macroblock is intra-coded if the number of intracoded macroblocks in the full-scale video is greater than 2. Moreover, in B frames, we should specify whether the downscaled macroblock is predicted from its previous or next frame. It may be bidirectional also. According to the scheme reported in [56], a downscaled macroblock is forward (backward) predicted if three or more of the four original macroblocks have forward (backward) prediction. Otherwise, it is bidirectionally predicted.

### 6.3.4 Downsizing MPEG2 Video

In [116] a DCT-based video downscaling system is presented. In Figure 6.8, the block diagram of the video downscaling system is shown. In this approach, the input video is VLC-decoded and inverse-quantized to get the DCT blocks. The interblocks are converted to intrablocks using the MBIMC scheme, and subsequently all the intrablocks are downscaled using a DCT domain image halving algorithm [36]. The AMVR technique [129] is used to compute the new motion vector for the downscaled video stream. Using this motion vector, residual errors for the downscaled video are computed in the DCT domain. For computing these errors, the predicted block is first formed using the MBIMC technique. Next, the type of the macroblock is determined using the technique (referred to as the MTSS in the block diagram) reported in [56]. Finally, from the inverse motion compensated macroblock, the corresponding predicted block is subtracted to provide the residual errors in the DCT domain. These DCT error blocks are further quantized and VLC encoded. If the type of downscaled macroblock is declared as *intra*, residual errors are not computed. For having control over the bit rate of the downscaled video, a scaling factor on quantization steps is also used.

In [129] a hybrid video downscaling system is proposed which performs a few operations in the spatial domain and others in the DCT domain. In this method, the input video is decoded completely and then downscaled in the spatial domain. However, the motion vector is derived from the motion vectors of the input video. To this end, the AMVR technique [129] is used. Using the derived motion vector, motion compensation for the downscaled video is performed in the spatial domain. Finally, the residual errors are encoded per MPEG standards. A hybrid approach to video downscaling reduces the

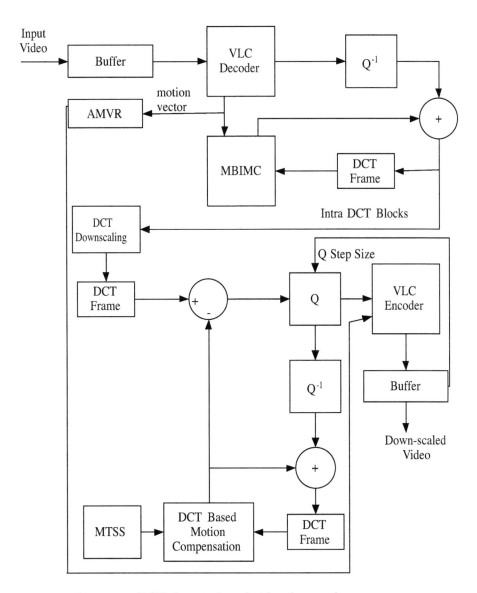

Figure 6.8: DCT domain based video downscaling system.

computation cost by a significant amount. However, as reported in [116], the pure DCT-based scheme provides better quality of downscaled video.

### 6.3.5   Arbitrary Video Downsizing

In [86], a hybrid scheme for arbitrary video downsizing is presented. Following the same approach of [129], in this case, motion vectors for downsized macroblocks are derived from the motion vectors of the full-scale video. Other operations related to downscaling are performed in the spatial domain. As the downsizing factor may not be an integer, motion vectors in the support area of the downsized macroblock are considered in the estimation. From the weighted average of these motion vectors, the downscaled motion vector is computed. In this case, the weights are not only computed from the activity of the corresponding macroblock, but also from its area of intersection with the support. Let $h_i \times w_i$ be the number of pixels in the area of the support within the $i$th macroblock with its motion vector as $\overrightarrow{mv_i}$. If the area of support consists of $n$ such macroblocks, the downscaled motion vector $\overrightarrow{mv_d}$ is estimated as follows:

$$\overrightarrow{mv_d} = \begin{bmatrix} \frac{1}{s_v} & 0 \\ 0 & \frac{1}{s_h} \end{bmatrix} \frac{\sum\limits_{i=1}^{n} \overrightarrow{mv_i} A_i B_i}{\sum\limits_{i=1}^{n} A_i B_i}, \tag{6.56}$$

where $s_h$ and $s_v$ are arbitrary downscaling factors in the horizontal and vertical directions. $A_i$ is the spatial activity of the $i$th macroblock, computed from its residual error and $B_i = h_i \times w_i$.

In [86], the same technique is applied to the H.263 encoded videos, and it is reported that introduction of weighting factors $B_i$'s improves the PSNRs of the downscaled video by around 2 dB compared to the scheme using the AMVR technique with its adaptation to the arbitrary downsizing.

Another example of a hybrid scheme for downscaling the HDTV video (of frame size 1920 × 1024) to the SDTV video (of frame size 720 × 384) is reported in [160]. In this case, both the television standards use MPEG-2 video compression, and motion vectors in the downscaled SDTV video are estimated from motion vectors of the corresponding macro blocks in the HDTV stream. The basic rate conversion is performed in the spatial domain. In this specific case, the area of support for the downscaled macroblock is found to be overlapping with an area of 4 × 4 adjacent macroblocks. Hence, the downscaled motion vector $\overrightarrow{mv_d}$ is estimated from a set of sixteen motion vectors, $\overrightarrow{mv_i}$, $1 \le i \le 16$. In this scheme, $\overrightarrow{mv_d}$ is given by the following expression:

$$\overrightarrow{mv_d} = \begin{bmatrix} \frac{1}{8} & 0 \\ 0 & \frac{1}{3} \end{bmatrix} WM(\{\overrightarrow{mv_i}\}), \tag{6.57}$$

where $WM(\{\overrightarrow{mv_i}\})$ is the weighted median of the set of motion vectors. Given the corresponding set of weights $W_i$, $1 \leq i \leq n$, the weighted median vector $\overrightarrow{mv_m}$ is computed as follows:

$$\overrightarrow{mv_m} = \arg \min_{\overrightarrow{mv_i} \in V} \left\{ W_i \sum_{j=1(j \neq i)}^{n} \| \overrightarrow{mv_i} - \overrightarrow{mv_j} \| \right\}.$$

## 6.4 Frame Skipping

Sometimes it is required to reduce the frame rate by dropping frames, resulting in lowering of the required bit rate of video transmission. Frame skipping may happen on the fly during the transmission. It may also be planned before transmission by judging the constraint imposed in a changing scenario. When a frame is dropped, reconstruction of macroblocks in subsequent frames referring to it gets affected, and it is required to compute new motion vectors for those macroblocks with respect to the nearby retained frame in the stream. Frame dropping may also occur over an interval. In that case, the cascading sequence of reconstruction needs to be taken care of. As mentioned earlier, the key concern of skipping a frame or a group of consecutive frames is to reestimate motion vectors of their adjacent frame. The computational problem is explained here with the help of an illustration in Figure 6.9. In Figure 6.9, the predicted block PB of a macroblock MB overlaps with four adjacent macroblocks numbered as 1, 2, 3, and 4 in the reference frame. Hence, the resulting motion vector of the current block should be determined by the set of motion vectors of those macroblocks in the reference frame ($\overrightarrow{v_1}$, $\overrightarrow{v_2}$, $\overrightarrow{v_3}$, and $\overrightarrow{v_4}$ in Figure 6.9). There are different approaches to performing this operation. A few of them are discussed below.

1. **Bilinear Interpolation [63]**: In [63] a motion vector corresponding to the target macroblock in the current frame is computed through bilinear interpolation of motion vectors accumulated in a cascading fashion from its referred set of macroblocks in the reverse sequence. However, the vector thus accumulated is further refined by searching the local minima of the SAD within a small window around it. The size of this window is determined by the number of skipped frames and the accumulated magnitudes of their motion vectors.

2. **Forward Dominant Vector Selection (FDVS) [166]**: In this method [166], a motion vector from the set of adjacent four macroblocks in the skipped frame is chosen in such a way that the predicted block of the current macroblock has the maximum overlap with the chosen macroblock. For example in Figure 6.9, suppose the predicted block PB has the maximum overlap with the macroblock 2. In that case, $\overrightarrow{v_2}$ is

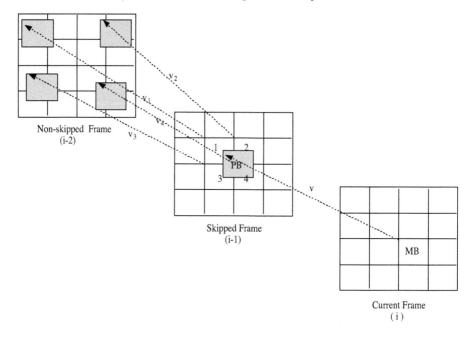

Figure 6.9: Motion Vector propagation from skipped frame.

chosen out of the four vectors and is used for determining the motion vector with respect to the $(i-2)$th frame. This process is repeated for each consecutively dropped frame until it encounters a nonskipped frame. However, for an increasing number of consecutive drops, the distance between the anchor frame and the current frame increases. In the process, there may be a mismatch in corresponding macroblock types, which have to be recomputed.

3. **Telescopic Vector Composition (TVC)[127]**: This technique follows a simple procedure of accumulating the resultant motion vector from all the motion vectors of the corresponding macroblocks of the dropped frames. A further refinement is searching local minima of the SAD over a small window around the location of the resultant vector. In this technique also, the macroblock type requires to be computed.

4. **Activity-Dominant Vector Selection (ADVS) [23]**: In this approach, the dominant vector is chosen based on the activity information of the macroblock in the reference frame. The activity information of a macroblock is computed by counting the number of nonzero quantized DCT coefficients of the residual error in the corresponding macroblock. We use other statistics also, such as the sum of the absolute values of

**Table 6.3**: Distinguishing features of main profiles of MPEG-2 and H.264/AVC

| Features | MPEG-2 | H.264/AVC |
|---|---|---|
| Motion estimation accuracy | $\frac{1}{2}$ pel | $\frac{1}{4}$ pel |
| Motion Vectors | Restricted by frame boundary | May occur outside the frame boundary |
| Number of Reference Frames | 1 for $P$ frames and 2 for $B$ frames | Up to 16 frames |
| Macroblock partitions | $16 \times 16$, $16 \times 8$ (interlace) | $16 \times 16$, $16 \times 8$, $8 \times 16$, $8 \times 8$, $8 \times 4$, $4 \times 8$, $4 \times 4$ |
| Intra frame spatial prediction types | None | 9 types |
| Transform used | Type-II $8 \times 8$ DCT | $4 \times 4$ ICT |
| Entropy encoding | VLC | CAVLC/CABAC |

DCT coefficients, etc. The advantage of this scheme compared to the FDVS is that it is faster to compute.

In [2], it is observed that the FDVS and the ADVS provide almost identical results. However, the latter has slight edge over fast-moving videos. The TVC is the fastest among all the above techniques, though its picture quality is a little poorer than the others. Most of the schemes use a hybrid approach for transcoding video with skipped frames by taking into account decoded frames in the pixel domain. However, in [46], a scheme is suggested which virtually performs a cascaded inverse motion compensation through the sequence of dropped frames in the DCT domain. Hence, the resulting macroblocks in the nonskipped frame becomes an intrablock. Though it has the disadvantage of losing coding efficiency, the encoding complexity reduces to a great extent by this process. The quality of the result also improves significantly.

## 6.5 Video Transcoding

In this section, we discuss transcoding techniques between two different compression standards, namely, the MPEG-2 and the H.264/AVC standards. In Table 6.3, the distinguishing features of the main profiles of these two compression standards are summarized [102].

Usually, the technique used for the transcoding is *cascaded pixel domain transcoding* (CPDT), where decoding of the source video stream is followed by its encoding by the different standard (see Figure 6.10). However, this process is aided by exploiting the relevant information already available in the compressed stream (see Figure 6.11). The latter technique is referred to as a hybrid approach to transcoding. However, there are also efforts to perform many tasks related to transcoding in the DCT/ICT domain itself. Hence, some of the issues and techniques for resolving them in video downsizing are also ap-

Figure 6.10: Cascading pixel domain transcoder.

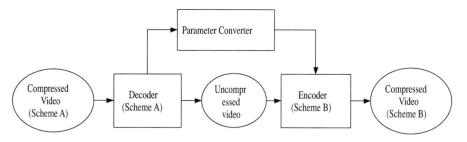

**Figure 6.11**: Transcoding aided by relevant information from the compressed stream.

plicable in this case. In this type of transcoding with heterogeneous standards, we need to perform *inverse motion compensation* (IMC), *macroblock type selection*, and *motion vector refinement* as elaborated in subsequent sections.

### 6.5.1   H.264 to MPEG-2

In [102], a hybrid transcoding scheme for converting an H.264 video into an MPEG-2 stream is presented. The architecture of this scheme is shown in Figure 6.12. As outlined in the hybrid transcoding approach of Figure 6.11, in this case also the H.264 video stream is decoded, and at the same time different information (or parameters) about macroblocks of inter frames is extracted and passed to the MPEG-2 encoder. The set of parameters extracted from the compressed stream includes the type of each frame, the type of each macroblock, and the set of motion vectors related to a partitioning of each macroblock. These parameters are mostly used to reduce the computation related to motion estimation in the MPEG-2 encoder (see to the MC block in the Encoder module of Figure 6.12). Besides in this approach, some of the macroblocks of intraframes are also directly converted into an intra-macroblock of the MPEG-2 stream in the transform domain following the conversion methods reported in [73]. In subsequent subsections, let us discuss a few specific issues related to the motion estimation and conversion of macroblocks of interframes and intraframes.

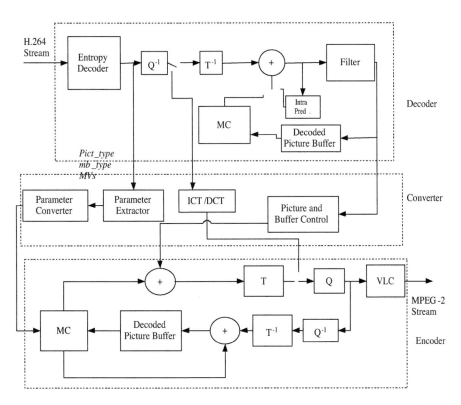

**Figure 6.12**: Hybrid transcoder from H.264 to MPEG-2 (adapted from [102]).

### 6.5.1.1    Motion Estimation

A macroblock of an interframe of H.264 standard is partitioned in different ways (see Section 1.5.3.5 of Chapter 1 and Table 6.3). However, in MPEG-2, a macroblock has only 1 or 2 motion vectors, depending on whether it belongs to a $P$ frame or a $B$ frame. For estimating these motion vectors in the transcoded video, motion vectors of different partitions of the H.264 macroblock are used. In fact, if we consider that all of them refer to the previous frame, a simple method of selecting one of them is to go for the vector resulting in the least amount of residual error (from the SAD computation). However, in the H.264 standard, the reference frame may not always be the previous or the next frame of the current one. Hence, we need to rescale these vectors if they point to a different frame (other than these two). Suppose a vector $\vec{v}$ refers to the $k$ backward frame. In that case, this vector is rescaled to $\frac{1}{k}\vec{v}$ to bring it within the domain of the previous frame only. To include the prediction from the next frame, we further consider the motion vector in its negative direction (i.e., $-\frac{1}{k}\vec{v}$ ). This concept is extended to handle motion vectors that point to regions outside the frame boundary. Moreover, H.264 works with *quarter pixel* accuracy in determining a motion vector. Hence, the corresponding vector is also scaled by 2 and rounded off to provide the motion vector with *half-pixel* resolution as specified in the MPEG-2 standard. By deriving a set of motion vectors using the above techniques from the H.264 macroblock, a vector is chosen for the MPEG-2 macroblock, that provides the least residual error. The SAD computation for these vectors is performed in the spatial domain as the decoded frames are available from the decoding module of the transcoder. In [102] a further search around these vectors within a small window (say, of size 3 × 3) is also carried out to improve the performance of the transcoder at the cost of marginal increase in computation. We may use the inverse motion compensation in the ICT domain for converting these macroblocks into intramodes as reported in [73].

### 6.5.1.2    Skip Macroblock

In H.264, skip macroblocks are differently treated from what is done in MPEG-2. For $P$ macroblocks, the skip indicates in H.264 that either the motion vector is zero or it remains the same as the corresponding block in the previous frame. In MPEG-2, it is considered a macroblock with zero motion vector. Hence, this becomes a special case of the H.264 skip definition for $P$ frames. However, for $B$ macroblocks, both the cases are included in the MPEG-2 standard. Hence, no extra processing is required for their conversion.

### 6.5.1.3    Intramacroblock

There are *nine* intraprediction modes in H.264 and out of them for three modes (*horizontal, vertical,* and *DC prediction* modes), compressed domain techniques for their direct conversion from ICT domain to the DCT domain are reported in [73] and [102]. These techniques also have some computational advantages with respect to spatial domain processing. However, for other modes, spatial domain processing is resorted as the equivalent computation becomes complex and inefficient in the transform domain. Moreover, with different types of partitioning in a macroblock, there could be $9^4 = 6561$ possible permutations of these intraprediction modes. Hence, pixel domain conversion is preferred in this case.

#### 6.5.1.4 Advantage of the Hybrid Approach

The foremost advantage of the hybrid approach is the savings in computation of motion vectors in the encoder, which usually covers 70% of its total computational cost. Hence, there is a saving of around 60% in the computation with almost no loss in the quality of transcoding with respect to the CPDT [102]. Moreover, when the estimate is further refined with a search around a small $3 \times 3$ window, an improvement of 0.3 dB is observed at the cost of a 5% increase in the computation cost.

### 6.5.2 MPEG-2 to H.264

Conversion of MPEG-2 video into an H.264 stream is simpler compared to its reverse operation, as most of the features of MPEG-2 are retained as special cases in H.264. Using the conversion from $8 \times 8$ DCT block to four $4 \times 4$ ICT blocks as discussed previously in this chapter (see Section 6.1.4), simple transcoding operations may be performed. However, as H.264 is more efficient in compressing videos, we would like to exploit its additional features in the transcoded stream. This requires different postprocessing operations even after straightforward conversion of frames. As an example, let us consider a scenario [161], where the intraframes are only transcoded from the MPEG-2 standards to the H.264 standards. An architecture of this scheme is shown in Figure 6.13. In this approach (see Figure 6.13), the mode for intrablock prediction for rate-distortion optimization (RDO) is decided in the transform domain. It has been shown in [161] that by using temporal correlation of mode decision in collocated blocks, we substantially reduce the computation with near-optimal performance of the transcoder.

## 6.6 Error Resilient Transcoding

Error resilience in an encoded stream is attributed to its tolerance and adaptability to the varying channel conditions of video transmission. It is difficult for an encoder to have prior knowledge about the characteristics of the communication channel, which may include multiple hops through routers and

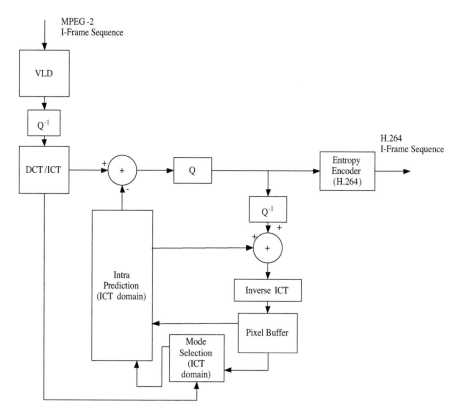

**Figure 6.13**: Transcoder for Intra H.264 to Intra MPEG-2 (adapted from [161]).

**Figure 6.14**: Error-resilient transcoding in video transmission (adapted from [145]).

switches. Moreover, today's world of network connectivity has become more heterogeneous. In particular, a range of client devices are operating over wireless channels. For delivering video content over these channels reliably, we should have error-resilient transcoding of video streams, acting as an interface between a low-loss network (such as a wired network) and a high-loss network (such as a wireless network). A schematic diagram of this interface is shown in Figure 6.14. There are different approaches to error-resilient coding. Some of them are listed below.

1. **Data partitioning and inserting more synchronization markers**: In [121], a scheme is proposed that modifies the encoding stream by introducing more slices and intrablocks with motion compensation in an encoded frame. The first one offers the synchronization points of spatial resiliency, and the latter provides the temporal resiliency. In [121], for reliable performance of the transcoder, the points of synchronization are adaptively chosen considering a model of error propagation in the decoder.

2. **Error-resilient entropy coding (EREC)**: In [140], error resiliency encoding for an MPEG-2 video stream is proposed by putting variable-length code blocks (VLC) into a fixed-length frame called EREC frame. Given $N$ code blocks of the video stream, there are $N$ EREC frames of fixed size, which is the average of the sizes of code blocks. Hence, blocks shorter than this length are well fitted within this frame. But longer blocks need to occupy additional spaces besides its exclusive EREC frame. These spaces are made available from the empty space of frames containing shorter blocks. The decoder also works on the same principle, given the knowledge of fixed frame size of EREC frames and rules of packaging of code blocks within them. In this case, the frame size should be included in the encoded stream as an additional overhead. However, the packaging rule simply follows the fitting of an individual block in its slotted frame from the starting index. If there is any empty space, rest of codes from successive larger blocks are packed inside. Hence encoded stream does not require any explicit additional information of this ordering. The advantage of this scheme lies in the fact that fixed length EREC frames act as synchronization units. Hence a corrupted frame is

dropped instead of all the frames between two synchronization markers in the stream.

3. **Reference frame selection (RFS)**: In this approach, given the knowledge of corrupted frame through a feed back channel, the transcoder adaptively shifts the reference of encoding of future frames. In that case, the error of reconstruction does not get propagated within two synchronization frames of a GOP. One simple technique for selecting the reference frame is to encode all the P frames from their nearest I frames, which are transmitted uncorrupted. However, this makes the process computationally intensive. In [19], these operations are carried out for macroblocks by updating their motion estimates and prediction errors following the similar strategy of frame skipping and inverse motion compensation.

4. **Adaptive intra refresh (AIR)**: In this approach [35], redundant intra-macroblocks are inserted in the video stream to arrest error propagation. However, this significantly increases the overhead of transmission. Hence, to regulate the bit rate, new quantization parameters are determined and used in the encoding.

---

## 6.7   Summary

Transcoding of images and videos is required due to heterogeneity of client devices and network connectivity. It also increases the interoperability of multimedia systems. A transcoder may produce the output stream in the same encoding standard of that of the input, by altering its spatial and/or temporal resolutions. This type of transcoding includes resizing of images and videos, skipping of frames of a video, etc. The other kind of transcoder is heterogeneous in character, which converts a stream from one standard to another. For example, we may transcode a JPEG2000-encoded image into a JPEG-compressed one. Similarly, we may convert an H.264 video into MPEG-2. Error-resilient transcoding is another type of transcoding, which operates both on the intermediate symbols (such as DCT coefficients) and the entropy-encoded code blocks. The objective of this type of transcoding is to provide a reliable and robust interface for transmission of media between a low-loss (such as wired network) and a high-loss (such as wireless network) network.

# Chapter 7

## Image and Video Analysis

In previous chapters, we discussed various image and video processing operations in the compressed domain such as filtering, color processing, resizing, and transcoding. These techniques and their variants are applied also in various other applications of analysis of images and videos in the compressed domain. They include video editing tasks such as document processing, caption localization, shot detection, etc. There are also different methods of indexing videos using features computed from the block DCT and the DWT space. Image and video digital watermarking and steganography in the compressed domain are also a major topic of research in the area of multimedia security. In this chapter, we review some of the representative approaches to compressed domain analysis of images and videos.

## 7.1   Image and Video Editing

We consider in this section different applications related to editing of images and videos in the compressed domain. To this end, let us review a few representative techniques of document processing, caption extraction, shot detection, and document merging in the transform domain.

## 7.1.1   Document Processing

In [33], a technique for processing JPEG compressed document images is presented. Various operations such as scaling, previewing, rotation, flipping, cropping, etc., are performed directly in the compressed domain. The interesting part of this work is that it also proposes algorithms for processing the entropy-encoded JPEG stream. Toward this, a measure on the cost of encoding a DCT block is used. The encoding cost is the number of bits allocated to the block in the stream. In this way, we obtain the encoding map (ECM) of an image as the distribution of encoding costs over the blocks. Given the ECM, it is possible to parse a video stream by extracting blocks or a set of adjacent blocks without performing entropy decoding. Hence, it is suggested in [33] to include the ECM, its row-sums, and decimated DC maps as side information in the compressed stream at the cost of a marginal increase in the size of the compressed data. However, we can also derive the ECM while decoding the entropy-encoded stream. The ECM provides important clues on activities of blocks. A smooth block without much detail has a low encoding cost, while occurrence of edges leads to higher values. Using the ECM, various operations such as cropping, segmentation, etc., are performed in the compressed domain. Typically, halftone regions in documents have high values in the ECM. Again, regions with text have higher concentration around their edges. Shaded regions have the sparse distribution of significant costs, and the background has very low encoding cost. However, the ECM alone is not sufficient for providing good segmentation. We should use other information such as DC coefficient of the block in addition to it for robust segmentation of different components of the document. For example, a high DC value with low encoding cost indicates that the block most likely belongs to the background. Availability of the ECM makes recompression of the processed image easier. It involves rearrangement and modification of entropy-encoded stream of blocks by readjustment of their DC offsets and end of block markers.

In [33], apart from the ECM, DCT coefficients are also used in the processing. For example, decimation and interpolation operations on images are carried out through block truncation and zero padding, respectively, as discussed in Chapter 5 on image resizing. DCT coefficients are also used for performing operations such as flipping and rotation by $90^{\circ}$, $180^{\circ}$, etc., of an individual block. Using the fast computational property of DCT for performing flipping as discussed in Eq. (3.27) in Chapter 3, these are easily performed. For the convenience of discussion, let us review the relationship here. A flipping matrix $\Phi_8$ is denoted as follows:

$$\Phi_8 = \begin{bmatrix} 0 & 0 & \ldots & 0 & 1 \\ 0 & 0 & \ldots & 1 & 0 \\ & & \ldots & & \\ 1 & 0 & \ldots & 0 & 0 \end{bmatrix}$$

Let $\Psi_8$ be the diagonal matrix given below.

$$\Psi_8 = \begin{bmatrix} 1 & 0 & 0 & \ldots & 0 & 0 \\ 0 & -1 & 0 & \ldots & 0 & 0 \\ 0 & 0 & 1 & \ldots & 0 & 0 \\ & & & \ldots & & \\ 0 & 0 & 0 & \ldots & 0 & -1 \end{bmatrix}$$

In that case, $C_8\Phi_8 = \Psi_8 C_8$ . From this relationship, we obtain the following.

**Theorem 7.1** *Let the rotated blocks of* $\mathbf{X}$ *by* $90°$, $-90°$, *and* $180°$ *be denoted as* $\mathbf{X}_{90}$, $\mathbf{X}_{-90}$, *and* $\mathbf{X}_{180}$, *respectively. Let* $\mathbf{X}_{\mathbf{VM}}$ *and* $\mathbf{X}_{\mathbf{HM}}$, *respectively denote its* vertical *and* horizontal *mirrors. In that case, they are given by the following operations:*

$$\begin{aligned} \mathbf{X}_{90} &= \Psi_8\mathbf{X}^T, \\ \mathbf{X}_{-90} &= \mathbf{X}^T\Psi_8, \\ \mathbf{X}_{180} &= \Psi_8\mathbf{X}, \\ \mathbf{X}_{\mathbf{HM}} &= \mathbf{X}\Psi_8, \\ \mathbf{X}_{\mathbf{VM}} &= \Psi_8\mathbf{X}. \end{aligned} \qquad (7.1)$$

□

We observe that the above computations can be performed without any multiplication. Even the magnitudes of the coefficients remain same. It is only the sign of a coefficient that may vary after transformation. Hence, it is possible to perform simple editing in the entropy-coded stream. Moreover, by exchanging the code segments using the ECM, it is possible to perform rotation and flipping operations over the whole image.

A typical example of document processing with a JPEG2000-compressed stream is also worth mentioning here. In [122], two regions of two different compressed images are merged by carrying out operations in entropy-coded streams. The technique exploits different synchronization markers of the code structure in the JPEG2000 stream, starting from *tiles*, *precincts*, and *code blocks* (see Section 1.4.2 of Chapter 1). JPEG2000 also defines quality layers for a precinct, which is divided into a set of packets, each containing an incremental contribution from each of its code blocks. Hence, a complete JPEG2000 code-stream is described as a concatenated list of packets, together with special marker segments mainly used for defining coding parameters. The ordering of these packets may vary depending on the needs of the application. In the technique presented in [122], structural elements (such as tiles, precincts, and codeblocks) are exchanged between two codestreams to merge a region of one image into another, without substantially affecting the image content within either region. In this approach, it is assumed that both the encodings are performed using the same parameters.

A precinct with its quality layers can be modeled as a pyramid whose bottom to top layers define multiresolution representation of wavelet coefficients for a region. Hence, for inserting a region from a source image to the destination one, we need to identify these pyramids from their precinct structure.

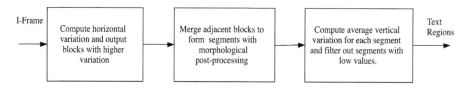

**Figure 7.1**: Block diagram for computation of caption localization from an I-Frame.

Then, the selected code structure is replaced from the source, resolution by resolution. This usually modifies the length of the codestream and hence we need to update information related to markers in it. However, in this technique, merging of regions is only possible along the boundaries of partitions in the images created during the encoding process.

### 7.1.2  Caption Localization in a Video

A few techniques have been put forward for localizing text in videos in the compressed domain. In [164], embedded captions are extracted from partially uncompressed MPEG video, by obtaining a low-resolution video from original MPEG sequences using either the DC components or the DC components plus two AC components of each block of I-frames (refer Eq. (4.14) of Chapter 4). By observing large interframe differences, text regions are detected as they indicate the appearance and disappearance of captions in video frames. However, the technique is limited by the fact that it is only capable of detecting captions that abruptly appeared or disappeared. In [49], another technique is proposed to detect text appearance events by counting the number of intracoded blocks in P or B-frames. This approach assumes that in the event of appearance and disappearance of captions, the corresponding blocks are usually intracoded. Due to its simplistic assumption, it is only applicable to the video segment which has a smooth shot transition between two I-frames.

In our discussion, we consider a texture-based caption localization technique proposed by Zhong et al. [169], which has more general applicability in detecting captions compared to the other two techniques discussed before. The technique is applicable to both JPEG images and MPEG videos as it deals with only I-frames of a video.

The technique proposed in [169] exploits the fact that there are DCT coefficients in the block that measure the directionality and periodicity of local image blocks. Using those coefficients, an $8 \times 8$ block is classified as *text* or *nontext* block. This preliminary selected set of candidate text blocks is further refined by performing postprocessing operations, such as morphological smoothing and cleaning, connected component analysis followed by filtering with more rigorous selection criteria. A block diagram of the technique is shown in Figure 7.1.

It is based on the intuitive observation that characters in a text line are responsible for a high response to the horizontal harmonics as they introduce rapid changes in intensity values. At the same time, a text region should also have a high response in vertical harmonics due to the occurrences of multiple text lines one after another in the vertical direction. These variations are measured by aggregating DCT coefficients associated with these harmonics. Consider that $X(i,j), 0 \le i, j, \le 7$, represent a DCT block in the luminance component. The horizontal variations and vertical variations are expressed as $E_h$ and $E_v$ respectively.

$$
\begin{aligned}
E_h &= \sum_{j=v_1}^{v_2} |X(0,j)|, \\
E_v &= \sum_{i=h_1}^{h_2} |X(i,0)|.
\end{aligned}
\tag{7.2}
$$

In [169], values of $v_1$, $v_2$, $h_1$, and $h_2$ are taken as 2, 6, 1, and 6, respectively. The thresholding for selecting the candidate text blocks in the first stage of computation (see Figure 7.1) is made adaptive by making it 1.45 times the average $E_h$ of all the blocks in the frame. Morphological operations such as *closing* and *opening* are carried out over the decimated image (considering unit blocks as points in the space) with a structuring element of size $1 \times 3$ for smoothing and eliminating spurious blocks, respectively. Then, connected components of adjacent blocks are formed. Finally, the segments with high average value of $E_v$ are accepted as text regions. On experimentation with 8 different videos consisting of 2360 I-frames, a recall rate of 99.17% with a false acceptance ratio of 1.87% was reported in [169]. The technique fails when the text font is large and characters are widely spaced, leading to low values of $E_h$ and $E_v$ in individual blocks.

### 7.1.3 Shot Detection

Shot detection is an important task required for analyzing videos. A shot is defined as the consecutive frames in a video that are captured with same camera view. A number of techniques are reported for detecting shots in the spatiotemporal domain of video. However, a few attempts are also made to perform this analysis in the compressed domain. For example, in [6], a technique for detecting cut is proposed by considering consecutive I-frames in the compressed stream. In this technique, for each frame a subset of DCT coefficients is selected as a representative vector and the normalized inner-product between vectors of two consecutive frames provide a measure of scene change. In [163], a technique is developed for shot detection by obtaining a low-resolution video from the DC coefficients of the blocks. Though DC coefficients for I-frames are easily obtained, we need to carry out inverse motion compensation to obtain DC values from P and B frames. Then any of the existing spatiotemporal shot detection techniques could be applied to this low-resolution video. A similar technique is also reported in [130]. In this case, cuts and gradual transitions are detected by comparing color histograms of consecutive frames of the low-resolution video.

Zhang et al. [167] used both motion vectors (MVs) and AC coefficients to measure the scene change in the block DCT space. In this technique, potential shot transitions are detected by a pairwise comparison of DCT coefficients of corresponding blocks of I frames. In the next stage, on examining the MVs of selected areas, the change is verified and the exact location of the cut is computed. However, the metric for cut detection does not work well with static frames.

In [85], direct binary edge maps are computed from the five lower-order AC coefficients[1] in blocks with the measurement of their strength and orientation. Next, histograms of these measurements are compared between two consecutive frames for detecting shot transitions.

In [77], shot detection is performed by taking into account macroblock coding modes and motion vectors in P and B frames. The task is accomplished in a two-pass scheme. In the first pass, potential shot boundaries are located in P frames with the help of rule-based modules. These are refined in the next pass using a neural network. The latter also becomes useful in distinguishing dissolves from object and camera motions, leading to further subdivision of a detected shot. For formation of rules, a specific sequence of picture types in a GOP is considered, namely, "IBBPBBPBBPBB".

In the first scan, temporal peaks in the number of intra-coded MBs in P frames are detected. They are good indicators of abrupt and gradual transitions. To find the exact location of the cuts, rules that check the number of backward and forward MBs in the context of two adjacent B frames are applied. For detecting the precise location of the *black fade*[2] boundaries, other sets of ordered rules using the number of interpolated, backward, and intra-coded MBs are used. Ends of fade-outs are determined from the number of interpolated and backward MBs within its 4 adjacent B frames. Similarly, beginnings of fade-ins are obtained by checking the number of backward, interpolated and intra MBs of up to 3 B and 1 P frames. In the second scan, a learning vector quantizer (LVQ) [76] is used to distinguish dissolves from object and camera movements and to precisely locate boundaries of the gradual transitions found by the rough scan.

---

[1]$(i, j)$th coefficient in the block such that $i + j \leq 2$.

[2]In a video fade, the sequence gradually fades to (or from) a single color, usually black or white. A black fade is gradual fading to (or from) a black frame.

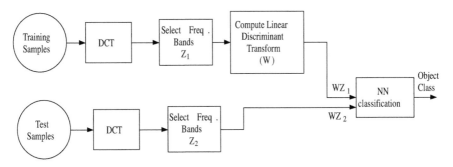

Figure 7.2: Block diagram for recognition of faces and palmprints.

## 7.2   Object Recognition

There have also been a few efforts in identifying events and objects in the compressed domain. In particular, quite a few techniques have been reported for detection and recognition of human faces and actions. They find their applications in biometry and video surveillance. Some of them are discussed in this section.

In [68], features from DCT frequency bands are selected using a separability measure for their applications in face and palmprint recognition with the help of a linear discriminant based classification. In this approach, the $k$th frequency band $R(k)$ of the DCT block $\mathbf{X}$ is defined as follows:

$$R(k) = \{\mathbf{X}(i,j) | max(i,j) = k\}. \tag{7.3}$$

Through separability analysis, a set of frequency bands is recommended for the discriminant analysis. In the next stage, using an improved Fisherface method [15], linear discriminant functions are obtained, which further reduces the dimension of the feature space. Finally, using nearest-neighbor classification the recognition of the object (face or palmprint as the case may be) takes place. The block diagram of this approach is shown in Figure 7.2. With this system, a high recognition rate on different databases ranging from 97.5% to 98.13% is reported.

In [150], the computation for detecting faces in I-frames of MPEG video is carried out in three stages. First, based on DC values of chrominance blocks, skin regions are identified. Due to $4 : 2 : 0$ sampling of components, each chrominance block corresponds to a $16 \times 16$ macroblock in the luminance component. Hence, the candidate face blocks are obtained in units of macroblocks, which are merged to form segments for a possible face region. Before merging, morphological operations are carried out to cover gaps and eliminate spurious blocks in the processing.

In the next stage, further pruning in the numbers of candidate regions

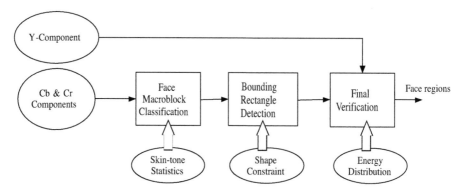

Figure 7.3: Block diagram for face detection.

takes place by considering allowable aspect ratios of bounding rectangles for human faces, which is taken as a value within the interval between 1 and 1.7. This provides considerable tolerance in the allowed values. There are also other constraints imposed in accepting a region, such as areas of the bounding rectangles. The accepted set of regions is further tested by a binary template matching technique. The templates are formed by considering possible combinations of aspect ratios and sizes.

In the final stage of this computation, candidate face regions are verified by examining the energy content in vertical and horizontal frequencies of the blocks. As the face region contains considerable details, it is expected that proportional amounts of these components should be quite high. The block diagram of this technique is shown in Figure7.3. The technique reported a detection accuracy ranging from 85% to 92% on different data sets.

In [28], however, verification order for face detection is reversed. Here, first gradient energy components in horizontal, vertical, and diagonal directions are computed following the same approach of aggregating pertinent DCT coefficients. These energy components form feature vectors and are used in a classifier for detecting candidate face macroblocks. In this regard, two types of classifiers are used. One is rule based, where relative proportions of different energy components determine the face or nonface regions. The other classifier is used as a neural network trained by labeled samples. In the final stage, candidate face macroblocks are verified from the chrominance components, which should belong to skin colors. In [28] it is reported that the technique achieved an overall detection rate of 75% at a precision of 83% using the rule-based approach. Using the neural network, the corresponding values were improved to 85% and 89%.

In [11], motion vectors of macroblocks for P and B frames are used to classify human actions such as walking, running, jumping, bending up, bending down, twisting right, and twisting left. The classifier used in this work is a *hidden Markov model* (HMM) classifier. Three different types of feature vectors are formed from the distribution of motion vectors, which are normalized in each frame with respect to nearest P frames. The GOP of the MPEG video in this work is considered in the form of "$IB_1B_2PB_1B_2\ldots$" with a length of 12 frames. As macroblocks of B frames have both forward and backward motion vectors, one of these groups is considered for a frame. For example, for $B_1$ frames, forward vectors are used, while for $B_2$, backward vectors are considered. Feature vectors are formed from these normalized motion vectors for a frame in the following way.

1. 1-D projection descriptor: In this case, only horizontal and vertical components of motion vectors are considered. The distribution of these components as a histogram over a set of finite intervals from the feature vectors.

2. 2-D polar descriptor: Motion vectors are represented in polar coordinates and distribution over the discrete cells in the polar coordinate space provide feature vectors.

3. 2-D Cartesian descriptor: Distribution of motion vectors in discrete cells of the cartesian coordinate space provides the third type of feature vector.

In the training phase, $k$-means clustering over the set of feature vectors for a known action is carried out, and each cluster generates a symbol for the given action, and the cluster representative represents the corresponding symbol. A feature vector from a frame is classified into one of these symbols using the nearest neighbor classifier and the sequence of these symbols provides the observation sequence for the HMM classifier, which is trained to detect an action and is used as a classifier subsequently. Using this technique, a recognition accuracy of more than 90% has been achieved for the set of actions mentioned before.

## 7.3  Image Registration

Image registration is the process of computing a global transformation between two images such that one of them after transformation gets aligned with the other. In the compressed domain, very few attempts have been made in this problem. They are also limited to handling image translation only. In [83], such a technique for computing translation between two images is reported. The technique deals with only DC coefficients, which form low-resolution images for both of them and then use edge-based cross-correlation measure to detect the translation, if any, between two images. As the low-resolution image is reduced by $\frac{1}{8}$th in each dimension, the translation offsets are multiplied by 8 to provide the values in the original image space. To provide the robustness in the process, four edge detectors along four different directions are applied. The binary maps produced by each of them are used for detecting the translation parameters providing the maximum cross-correlation measure in the image space. Finally, the one with the maximum score (out of the four derived offsets) is chosen for image registration.

A technique for computing registration between two images is presented in [64], which uses the *sign only correlation* (SOC) between two images in the DCT space. The concept of the SOC is extended from the *phase-only correlation* (POC), which is defined in the DFT domain. The POC has the advantage of estimating transformation parameters related to translation, rotation and scaling with subpixel accuracy. In [64], the relationship between the SOC and the POC has been established for using the former in estimating the translational displacement between two images in the DCT space. Let us discuss this useful relationship.

Given a sequence $x(n)$, consider its DFT as $X_f(k)$ and its type-II DCT as $X_c(k)$. We represent $X_f(k)$, a complex number, with its magnitude and phase term as $X_f(k) = |X_f(k)|X_f'(k)$. On the other hand $X_c(k)$ being a real number, its phase term is represented by its sign $(+/-)$ and in the same notation factorize $X_c(k)$ as $|X_c(k)|X_c'(k)$. Note that the value of nonzero $X_c'(k)$ could be either 1 or $-1$.

**Definition 7.1** *Given two sequences $x(n)$ and $y(n)$, their cross-spectrum $R_f(K)$ in the DFT domain is defined as follows:*

$$R_f(K) = X_f'(k)(Y_f'(k))^*, \tag{7.4}$$

*where $(A)^*$ denotes the complex conjugate of $A$.* ◻

The inverse transform of the cross-spectrum provides the *phase only correlation* (POC) between two sequences. Its definition is given below.

**Definition 7.2** *The POC of two sequences $x(n)$ and $y(n)$ is defined as*

$$r_f(n) = \frac{1}{N} \sum_{k=0}^{N-1} R_f(K)e^{j2\pi \frac{k}{N} n}, \tag{7.5}$$

*where $N$ is the length of the sequences.* ◻

The DCT *sign-only correlation* (SOC) is similarly defined from the cross-spectrum of DCT phases $(+1, -1, \text{or } 0)$. The inverse operation over this cross-spectrum is a type-I DCT inverse. The precise definition is given below.

**Definition 7.3** *Given two sequences $x(n)$ and $y(n)$ of length $N$, its cross-spectrum $R_c(K)$ in the DCT domain is defined as follows.*

$$R_c(K) = X_c'(k)Y_c'(k). \tag{7.6}$$

*Then the SOC between $x(n)$ and $y(n)$ is defined as*

$$r_c(n) = \sum_{k=0}^{N} \alpha(k) R_c(K) \cos(\frac{\pi kn}{N}), \qquad (7.7)$$

*where $R_c(N) = 0$ and $\alpha(k)$ is $\frac{1}{2}$ for $k = 0$ and $k = N$. For other values of $k$, $\alpha(k)$ is 1.* $\square$

It is shown in [64] that the DCT SOC is the same as the POC from the DFT of symmetric extensions of two sequences. The symmetric extension corresponds to the type-II DCT of the sequence, which means it is an 'HSHS' type of extension (see Section 2.2.3.2 of Chapter 2). Hence, the peak in $r_c(n)$ provides the necessary translational parameter for registering $x(n)$ with respect to $y(n)$.

In some work [7, 8, 16], a similarity measure $S_c$ based on the SOC is used for image matching and retrieval as defined below.

$$S_c = \frac{1}{K} \sum_{k=0}^{N-1} R_c(k), \qquad (7.8)$$

where $K$ is a constant. It can be shown that for $K = N$, $S_c$ is the same as $r_c(0)$. Hence, the SOC can be used for finding the similarity between two images in the DCT domain.

---

## 7.4 Digital Watermarking

A digital watermark is a kind of digital signature, which is embedded in a multimedia object for attaching the ownership of the signatory. Watermarking techniques have become useful in enforcing copyright protection of a document, so that the embedded signature is detected or extracted later for the purpose of authentication of the source. A few techniques [22, 58, 89] are also reported in the compressed domain for watermarking compressed documents. We present here a technique [89] that embeds a watermarking bit in mid-frequency DCT coefficients of a block. The middle of the frequency band is chosen as even a slight modification to low-frequency coefficients causes perceptible loss of fidelity of the image. On the other hand, due to compression and other operations, high-frequency components are liable to get eliminated. This destroys the watermarking signatures from the image.

Let $\mathbf{X_i}$ be the $i$th $8 \times 8$ DCT block. Let us consider the serialization of the frequency components in such a way that selected bands of frequencies are represented consecutively in the sequence $f(k) \in \mathbf{X_i}, 0 \le k \le 63$. Say the selected band occurs between $k_0$ and $k_1$ such that $k_0 < k_1$. Let $u(i)$ define the mean values of this set of coefficients such that $u(i) = \frac{1}{k_1-k_0+1} \sum_{k=k_0}^{k_1} f(k)$.

Let us also represent the mean of $u(i)$s as $\overline{u}$. The technique modulates $f(i)$ in such a way that in the embedded stream, their mean becomes greater than modified $\overline{u}$ if the $i$th watermarking bit $w_i$ is 1. Otherwise, the modified $u(i)$ will be less than $\overline{u}$. It has been shown in [89] that this is accomplished with the following processing of the $u(i)$'s in the embedded stream. Let $u(i)$ be modified to $u^h(i)$ as given below.

$$u^h(i) = u(i) + d^h(i) \qquad (7.9)$$

where

$$d^h(i) = \begin{cases} +|Q(i)| - (u(i) - \overline{u}) & \text{if } w(i) > 0 \\ -|Q(i)| - (u(i) - \overline{u}) & \text{if } w(i) < 0 \end{cases} \qquad (7.10)$$

In the above, $Q(i)$ is flexibly chosen for the $i$th block and usually kept at a constant value for all the blocks. To bring the above changes in $u(i)$, each $f(k)$ of the $i$th block between $k_0$ and $k_1$ is changed in the same way. The technique is reported to be robust against various attacks for destroying watermarks.

---

## 7.5 Steganography

Steganography is the art of hiding data by embedding messages in innocuous looking objects, such as digital images. The embedding process modifies the original (cover) image and turns it into a stego image. However, to an outsider, the image appears as normal and natural as any other in the set. Only the concerned communicating ends share a secret stego key to decipher the message from the stego image. A steganographic process should ensure the statistical undetectability of the hidden data given the complete knowledge of the embedding mechanism and the source of cover objects but not the stego key [115]. Various techniques [9, 22, 118] have been put forward for performing steganography in the compressed domain. A few representative ones are briefly discussed here.

The JSteg [9] is one of the initial implementations of the steganography in the JPEG domain, where the least significant bits of quantized DCT coefficients are used for embedding. However, it excludes values such as 0 and 1, which cause distortion in the image to a great extent. As the JSteg introduces characteristics artifacts into the histogram of DCT coefficients, it is easily detectable by histogram-based attacks [154]. To overcome this bottleneck, techniques preserving statistical properties of cover images are also advanced. For example, Provos [118] proposed an algorithm that first identifies the redundant DCT coefficients that have a minimal effect on the cover image, and then chooses bits in which it embeds the message.

In [138], DCT coefficients in the cover image are partitioned into two sets. One partition is used for data embedding, and the other one is kept for statistical restoration. To this end, a statistical restoration method is proposed, so that histogram modification due to alterations in the embedded set is compensated by changing coefficients in the restoration set. One of the constraint of this operation is to keep the *mean square error* (MSE) at a minimum while modifying the histogram. To ensure this, all the bins of the target histogram are compensated in an increasing order by mapping the input data with values in the same order. Another technique for steganography in the DCT domain is through *quantization index modulation* (QIM) [22]. In QIM [22], the choice of quantizers varies depending on stego message bits.

## 7.6 Image and Video Indexing

Indexing images and videos is the task of attributing them to a reference key, and useful for their storage and retrieval operations in a database management system. This key reference could be in the form of a feature descriptor, textual annotation, or representative part of the data. In the transform domain, there are quite a few works reported for indexing images that exploit their representation in the alternative form, and use them in forming feature vectors. In the following, we discuss a few of them. Subsequently, we have also considered video indexing that primarily focuses on its segmentation into shots and detecting key frames in a shot.

### 7.6.1 Image Indexing

In [137], initially an image is represented by a feature vector of dimension 32 from its $4 \times 4$ block DCT representation. The fields of the vector are formed by the means and variances of DCT coefficients of each spatial frequency (16 in number) across the image. The variance and the mean absolute values of each of these coefficients are computed over the entire image. Further, using Fisher discriminant analysis (FDA), the dimensionality of the feature vector is reduced. The FDA provides a linear combination of original components such that on average the resulting fields are maximally separated among training classes. The feature vector after reduction of dimensionality is used for indexing. Reeves et al. [120] used a similar concept in obtaining a feature descriptor. However, they operated on the $8 \times 8$ block DCT space. In this case, the feature vector is formed using variances of the first eight AC coefficients. This approach is computationally less intensive than the former approach.

In [133], an image is represented by a binary string of length 16 consisting of 0 and 1 only. The technique considers a random selection of 16 pairs of windows (multiples of $8 \times 8$ blocks in the JPEG standards) in a random sequence and compares each pair with their feature descriptors. A similar pair appends 0 in the string, otherwise it appends 1. The feature descriptor is generated from the average of DCT coefficients of blocks belonging to a window. This provides 64 features for a window. If the corresponding feature values of the two are very close (within a threshold), the pair generates 0 in the key string; otherwise it is 1. While searching over a set of images, each target image is converted into the key binary string of length 16 following the same sequence of comparison of pairs of windows at the same locations of the query image. The Hamming distance between the two provides the measure of similarity.

There are efforts in the DWT domain as well. In [21], a texture analysis scheme is presented through irregular tree decomposition, where similarity is computed from the moderate-resolution subband coefficients. In this scheme, the $K$ most important subbands (based on their energy) are used to form a $K$-dimensional feature vector.

## 7.6.2   Video Indexing

As mentioned, video indexing mostly involves selection of the key frames in a shot. Some of the shot detection algorithms have already been discussed before. Here, we discuss a few representative techniques related to key frame extraction.

### 7.6.2.1   Key Frame Selection

Like shot detection, for extracting key frames also, different techniques use information related to texture coding modes (intra, inter, etc.), the motion vectors, and the significant changes in DC coefficients, etc., from the compressed video.

In [10], a feature vector descriptor is used for representing a frame of a shot by taking into account its color and motion information. Prior to this, the video is converted into a low-resolution video by representing every block of its frames by their DC coefficients. The technique for obtaining such a video follows from [163]. There are several computational steps involved in the process of this conversion. First, the shot is segmented using color and motion information. Next, a fuzzy classification scheme is used to convert the frame with a feature vector, where each field of the vector denotes an accumulated sum of membership function to different color and motion classes (defined a priori). In the final stage of computation, key frames are extracted following two approaches. In one approach, the temporal trajectory of the feature vector in a multidimensional space is analyzed. The points of discontinuities of the curve (in this case, second-order discontinuities) provide the key frame locations. However, this generates too many key frames. Hence, in another approach, using a cross-correlation measure between a pair of feature vectors, an optimal set of a fixed number (known or assumed a priori) of key frames is

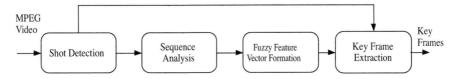

Figure 7.4: Block diagram for selection of key frames.

selected, which minimizes the average cross-correlation of all possible pairs in the set. The optimization process is further carried out through a logarithmic search or by a genetic algorithm (GA)-based technique. The block diagram of the extraction process is shown in Figure 7.4.

### 7.6.2.2 Key Video Object Plane Selection

In the MPEG-4 compression standard, representation of video is object-based. In this case, individual video objects (VOs) are encoded into separate bit streams (see Section 1.5.2 of Chapter 1). A temporal instance of a video object is called a video object plane (VOP). Similar to key frames of an MPEG-2 video, in MPEG-4 key VOPs are useful for visual summarization of the video object content in an object-based framework. Hence, instead of looking for key frames we would be interested in searching for the *key video object plane*. In [40], a technique for selecting the key video object plane using the shape information from an MPEG-4 compressed video is presented.

Encoding of VOPs is quite similar to the coding of a sequence of frames in MPEG-2, as discussed in Chapter 1. In this case, each I-VOP is encoded independently following the principles of encoding of textures of I frames similar to those of MPEG-2. For P and B VOPs, motion compensation and predictive coding are used. A binary mask fitted into a bounding rectangle of the object describes the shape of a VOP. The mask is known as the binary alpha plane, which defines whether a pixel belongs to the object or not. These data are also transmitted along with the usual shape information in encoded macroblocks corresponding to the same area. In [40], the shape of the video object (VO) is represented by the shape-coding modes of I, P, and B video object planes, as it avoids decoding of the shape information in the MPEG-4 bit stream. There are three shape-coding modes possible for IVOPs, namely, 1) transparent, if the block is empty; 2) opaque, if it is full; and 3) intra, if the block lies at the boundary of the object. In fact, in the compressed stream, only intrashape coding information is retained as others could be derived by analyzing the shape of the boundary blocks. The intrashape coding may be lossy also.

On the other hand, for P and B VOPs, there exist seven shape-coding modes according to the MPEG-4 standard. In addition to the transparent, opaque, and intra-coding modes, there are also four inter-shape coding modes involving the transmission of motion vectors and prediction information. It is only when the quality of prediction is not good, that the intrashape coding mode is employed for them. Characterization of a block to one of three intra shape modes is possible only when the VOP is decoded and the topological relationship of the block with respect to the object is determined. However, to save computation, in [40] these modes are predicted based on a set of rules operating with information available from intrashape modes of the same location from temporally nearer frames. Finally, on assigning each block a value corresponding to its intrashape coding mode, a feature vector is formed to represent the VOP.

In the beginning the first VOP of the stream is considered the *key* VOP. Subsequently, following the appearance of VOPs in the temporal order, a distance between the *key VOP* and the VOP newly appeared is measured using modified forms of Hamming and Hausdorff distance [52] functions. If the distance is significant (that is, greater than a threshold), the new VOP is replaced as the *Key VOP*. In this way, the sequence of *key VOP*s are computed.

---

## 7.7   Summary

In this concluding chapter, a brief treatment of different types of analysis of images and videos in the compressed domain is presented. Representative techniques of different video editing operations, object recognition, image registration, digital watermarking and steganography, and image and video indexing are discussed. This list is not all-inclusive. Analysis and processing of images and videos in the compressed domain are required for accomplishing many other tasks, with the objectives of saving computational time and storage, as well as improving quality of results. The present study only highlights some of the fundamental aspects and key concepts behind the development of such techniques.

# Bibliography

[1] S. Aghagolzadeh and O.K. Ersoy. Transform image enhancement. *Opt. Eng.*, 31:614–626, Mar. 1992.

[2] I. Ahmad, X. Wei, Y. Sun, and Y.Q. Zhang. Video transcoding: An overview of various techniques and research issues. *IEEE Trans. Multimedia*, 7(5):793–804, Oct. 2005.

[3] N. Ahmed, N. Natarajan, and K.R. Rao. Discrete cosine transform. *IEEE Trans. Comput.*, pages 90–92, Jan. 1974.

[4] M. Antonini, M. Barlaud, P. Mathieu, and I. Daubechies. Image coding using wavelet transform. *IEEE Trans. Image Process.*, 1(2):205–220, Apr. 1992.

[5] Y. Arai, T. Agui, and M. Nakajima. A fast DCT-SQ scheme for images. *Trans. IEICE*, E71:1095–1097, Nov. 1988.

[6] F. Arman, A. Hsu, and M.-Y. Chiu. Image processing on encoded video sequences. *Multimedia Syst.*, 1:211–219, 1994.

[7] F. Arnia, I. Iizuka, M. Fujiyoshi, and H. Kiya. Compressed domain action classification using HMM. *IEICE Trans. Fundamentals*, E89-A(6):1585–1593, June, 2006.

[8] F. Arnia, I. Iizuka, M. Fujiyoshi, and H. Kiya. Fast and robust identification methods for JPEG images with various compression ratios. In *Proc. IEEE Int. Conf. Speech, Acoust. Signal Process.*, pages II–397–II–400, Toulouse, France, May, 2006.

[9] I. Avcibas, M. Kharrazi, N.D. Memon, and B. Sankur. Image steganalysis with binary similarity measures. *EURASIP J. Appl. Signal Process.*, 17:2749–2757, 2005.

[10] Y.S. Avrithis, A.D. Doulamis, N.D. Doulamis, and S.D. Kollias. A stochastic framework for optimal key frame extraction from MPEG video databases. *Comput. Vis. Image Understanding*, 75(1/2):3–24, July/Aug. 1999.

[11] R.V. Babu, B. Anantharaman, K.R. Ramakrishnan, and S.H. Srinivasan. Compressed domain action classification using HMM. *Pattern Recognition Lett.*, 23:1203–1213, 2002.

[12] K. Barnard, V. Cardei, and B. Funt. A comparison of computational color constancy algorithms- PART I: Methodology and experiments with synthesized data. *IEEE Trans. Image Process.*, 11(9):972–984, Sept. 2002.

[13] K. Barnard, V. Cardei, and B. Funt. A comparison of computational color constancy algorithms- PART II: Experiments with image data. *IEEE Trans. Image Process.*, 11(9):985–996, Sept. 2002.

[14] K. Barnard, L. Martin, B. Funt, and A. Coath. A data set for color research. *Color Res. Appl.*, 27(3):148–152, 2000.

[15] P. N. Belhumeur, J. P. Hespanha, and D. J. Kriegman. Eigenfaces vs. fisherface: Recognition using class specific linear projection. *IEEE Trans. Pattern Analysis Machine Intelligence*, 19(7):711–720, July, 1997.

[16] J. Bracamonte, M. Ansorge, F. Pellandini, and P. A. Farine. Efficient compressed domain target image search and retrieval. In *Proc. 4th Int. Conf. Image Video Retrieval*, pages 154–163, Singapore, July 20-22, 2005.

[17] G. Buchsbaum. A spatial processor model for object colour perception. *J. Franklin Inst.*, 310:1–26, 1980.

[18] W.K. Cham. Development of integer cosine transforms by the principle of dyadic symmetry. *IEE Proc., Part I*, 136(4):276–282, Aug. 1989.

[19] Y.L Chan, H.K. Cheung, and W.C. Siu. Compressed-domain techniques for error-resilient video transcoding using RPS. *IEEE Trans. Image Process.*, 18(2):357–369, Feb. 2009.

[20] S. F. Chang and D. G. Messerschmitt. Manipulation and compositing of MC-DCT compressed video. *IEEE J. Sel. Areas Commun.*, 13(1):1–11, Jan. 1995.

[21] T. Chang and C.C.J. Kuo. Texture analysis and classification with treestructured wavelet transform. *IEEE Trans. Image Process.*, 2(4):429–441, Oct. 1993.

[22] B. Chen and G.W. Wornell. Quantization index modulation: A class of provably good methods for digital watermarking and information embedding. *IEEE Trans. Information Theory*, 47(4):1423–1443, May, 2001.

[23] M.-J. Chen, M.-C. Chu, and C.-W. Pan. Efficient motion-estimation algorithm for reduced frame-rate video transcoder. *IEEE Trans. Circuits Syst. for Video Technol.*, 12(4):269–275, Apr. 2002.

[24] W.-H. Chen, C.H. Smith, and S.C. Fralick. A fast algorithm for the discrete cosine transform. *IEEE Trans. Commun.*, COM25(9):1004–1009, Sept. 1977.

[25] N.I. Cho and S.U. Lee. A fast $4 \times 4$ DCT algorithm for the recursive 2-D DCT. *IEEE Trans. Signal Process.*, 40(9):2166–2173, Sept. 1992.

[26] N.I. Cho, I.D. Yun, and S.U. Lee. On the regular structure for the fast 2-D DCT algorithm. *IEEE Trans. Circuits Syst. II: Analog Digital Signal Process.*, 40(4):259–266, Apr. 1993.

[27] C. Christopoulos, A. Skodras, and T. Ebrahimi. The JPEG2000 still image coding system: an overview. *IEEE Trans. Consumer Electron.*, 46(4):1103–1127, Nov. 2000.

[28] T.-S Chua, Y. Zhao, and M.S. Kankanhalli. Detection of human faces in a compressed domain for video stratification. *Vis. Comput.*, 18:121–133, 2002.

[29] CIE. Colorimetry. *Central Bureau CIE*, 15(2), 1986.

[30] I.J. Cox, M.L. Miller, J.A. Bloom, J. Fridrich, and T. Kalker. *Digital Watermarking and Steganography*. Morgan Kaufmann, Burlington, MA, 2nd edition, 2008.

[31] I. Daubechies. *Ten Lectures on Wavelets.*, volume 61. CBMS-NSF Regional Conf. Series in Appl. Math. Society for Industrial and Appl. Mathematics, Philadelphia, PA, 2004.

[32] I. Daubechies and W. Sweldens. Factoring wavelet transforms into lifting steps. *J. Fourier Anal. Appl.*, 4(3):247–268, 1998.

[33] R.L. de Queiroz. Processing JPEG-compressed images and documents. *IEEE Trans. Image Process.*, 7(12):1661–1672, Dec. 1998.

[34] S. Dewitte and J. Cornelis. Lossless integer wavelet transform. *IEEE Signal Process. Lett.*, 4(6):158–160, 1997.

[35] S. Dogan, A. Cellatoglu, M. Uyguroglu, A. H. Sadka, and A. M. Kondoz. Error-resilient video transcoding for robust internetwork communications using GPRS. *IEEE Trans. Circuits Syst. Video Technol.*, 12(6):453–464, Jun. 2002.

[36] R. Dugad and N. Ahuja. A fast scheme for image size change in the compressed domain. *IEEE Trans. Circuits Syst. for Video Technol.*, 11(4):461–474, 2001.

[37] P. Duhamel and C. Guillemot. Polynomial transform computation of the 2-D DCT. In *Proc. IEEE Int. Conf. Acoustics, Speech and Signal Process.*, volume 3, pages 1515–1518, Albuquerque, New Mexico, Apr. 3–6, 1990.

[38] P. Duhamel and H. H. Mida. New $2^n$ DCT algorithms suitable for VLSI implementation. In *Proc. IEEE Int. Conf. Acoustics, Speech, and Signal Process.*, volume 12, pages 1805–1808, Dallas, TX, Apr. 6–9, 1987.

[39] M. Ebner, G. Tischler, and J. Albert. Integrating color constancy into JPEG2000. *IEEE Trans. Image Process.*, 16(11):2697–2706, Nov. 2007.

[40] B. Erol and F. Kossentini. Automatic key video object plane selection using the shape information in the MPEG-4 compressed domain. *IEEE Trans. Multimedia*, 2(2):129–138, June, 2000.

[41] E. Feig and S. Winograd. Fast algorithms for the discrete cosine transform. *IEEE Trans. Signal Process.*, 40(9):2174–2193, Sept. 1992.

[42] G.D. Finlayson. Color in perspective. *IEEE Trans. Pattern Anal. Machine Intelligence*, 18(10):1034–1038, Oct. 1996.

[43] G.D. Finlayson, S.D. Hordley, and P.M.Hubel. Color by correlation: a simple, unifying framework for color constancy. *IEEE Trans. Pattern Anal. Machine Intelligence*, 23(11):1209–1221, Nov. 2001.

[44] D.A. Forsyth. A novel algorithm for color constancy. *Int. J. Comput. Vis.*, 5(1):5–36, 1990.

[45] M.W. Frazier. *An introduction to wavelets through linear algebra.* Springer-Verlag, New York, 1999.

[46] K.T. Fung, Y.L. Chan, and W.C. Siu. New architecture for dynamic frame-skipping transcoder. *IEEE Trans. Image Process.*, 11(8):886–900, Aug. 2002.

[47] D. Le Gall. MPEG : A video compression standard for multimedia applications. *Commun. ACM*, 34(4):47–58, Apr. 1991.

[48] D. Le Gall and A. Tabatabai. Subband coding of digital images using symmetric kernel filters and arithmetic coding techniques. In *Proc. IEEE Int. Conf. Acoustics, Speech Signal Process.*, pages 761–764, New York, Apr. 3–6, 1988.

[49] U. Gargi, S. Antani, and R. Kasturi. Indexing text events in digital video databases. In *Proc. 14th Int. Conf. Pattern Recognition*, pages 916–918, Brisbane, Australia, Aug. 17–20, 1998.

[50] R. Gershon, A.D. Jepson, and J.K. Tsotsos. From [R,G,B] to surface reflectance: computing color constant descriptors in images. *Perception*, pages 755–758, 1988.

[51] B. Girod, A.M. Aaron, S. Rane, and D. Rebollo-Monedero. Distributed video coding. *Proc. IEEE*, 93(1):71–83, Jan. 2005.

[52] R.C. Gonzales and R. E. Woods. *Digital Image Processing.* Addison-Wesley, Reading, MA, 1992.

[53] R. C. Gonzalez and R. E. Woods. *Digital Image Processing using MATLAB®.* Pearson Education, Indian Reprint 1st edition, 2004.

[54] H.G. Grassmann. Theory of compound colors. *Philos. Mag.*, 4(7):254–264, 1854.

[55] E.A. Guillemin. *The Mathematics of Circuit Analysis.* Oxford and IBH Publishing Co., (Indian Ed.), Calcutta, India, 1967.

[56] M. R. Hashemi, L. Winger, and S. Panchanathan. Macroblock type selection for compressed domain downscaling of MPEG video. In *Proc. IEEE Canadian Conf. Electrical and Computer Engineering*, volume 4, pages 35–38, Kobe, Japan, Oct. 26–30, 1999.

[57] M.R. Hashemi, L. Winger, and S. Panchanathan. Compressed domain motion vector resampling for downscaling of mpeg video. In *Proc. IEEE Int. Conf. Image Process.*, volume 4, pages 276–279, Kobe, Japan, Oct. 26–30, 1999.

[58] J.R. Hernndez, M. Amado, and F.P. Gonzlez. DCT-domain watermarking techniques for still images: Detector performance analysis and a new structure. *IEEE Trans. Image Process.*, 9(1):55–68, Aug. 2000.

[59] H. Hou. A fast recursive algorithm for computing the discrete cosine transform. *IEEE Trans. Acoustics, Speech and Signal Process.*, 35(10):1455–1461, Oct. 1987.

[60] Q. Hu and S. Panchanathan. Image/video spatial scalability in compressed domain. *IEEE Trans. Ind. Electron.*, 45(Feb.):23–31, 1998.

[61] R.W.G. Hunt. *Measuring Colour.* Ellis Horwood Ltd. Publ., Chichester, UK, 2nd edition, 1987.

[62] Y.S. Park H.W. Park and S.K. Oh. L/M-image folding in block DCT domain using symmetric convolution. *IEEE Trans. Image Process.*, 12(9):1016–1034, Sept. 2003.

[63] J.-N. Hwang and T.-D. Wu. Motion vector re-estimation and dynamic frame-skipping for video transcoding. In *Conf. Rec. 32nd Asilomar Conf. Signals, System and Comput.*, volume 2, pages 1606–1610, Pacific Grove, CA, Nov. 2–5, 1998.

[64] I. Ito and H. Kiya. DCT sign-only correlation with application to image matching and the relationship with phase-only correlation. In *Proc. IEEE Int. Conf. Speech, Acoust. Signal Process.*, pages I–1237–I–1240, Honolulu, Hawai, Apr. 15–20, 2007.

[65] A. Jacquin. Image coding based on a fractal theory of iterated contractive image transformations. *IEEE Trans. Image Process.*, 1(1):18–30, 1992.

[66] Seong Hwan Jang and Nikil Jayant. An adaptive non linear motion vector resampling algorithm for downscaling video transcoding. In *Proc. IEEE Int. Conf. Multimedia and Expo*, volume 2, pages 229–232, Baltimore, MD, July 6–9, 2003.

[67] J. Jiang and G. Feng. The spatial relationships of DCT coefficients between a block and its sub-blocks. *IEEE Trans. Signal Process.*, 50(5):1160–1169, May, 2002.

[68] X.-Y Jing and D. Zhang. A face and palmprint recognition approach based on discriminant DCT feature extraction. *IEEE Trans. Syst. Man and Cybernatics Part B: Cybernatics*, 34(6):2405–2415, Dec. 2004.

[69] D. J. Jobson, Z. Rahman, and G. A. Woodell. Properties and performance of a center/surround retinex. *IEEE Trans. Image Process.*, 6(3):451–462, Mar. 1997.

[70] N. F. Johnson and S. Jajodia. Steganography: Seeing the unseen. *IEEE Comput.*, (2):26–34, Feb. 1998.

[71] D.B. Judd. Reduction of data on mixture of color stimuli. *Bur. Standard J. Res.*, 4:515–548, 1930.

[72] S.-H. Jung, S.K. Mitra, and D. Mukherjee. Subband DCT: Definition, analysis and applications. *IEEE Trans. Circuits Syst. for Video Technol.*, 6(3):273–286, June, 1996.

[73] T. Kalyani, A. Bhartiya, V. Patil, R. Kumar, and J. Mukherjee. DCT domain transcoding of H.264/AVC video to MPEG-2 video. In *Proc. 5th Indian Conf. Comput. Vis. Graphics Image Process.*, pages 696–707, Madurai, India, Dec. 14–16, 2006. LNCS-4338.

[74] F. A. Kamangar and K. R. Rao. Fast algorithms for the 2-D discrete cosine transform. *IEEE Trans. Comput.*, C-31(9):899–906, Sept. 1982.

[75] M. Karczewicz and R. Kurceren. The SP- and SI-Frames design for H.264/AVC. *IEEE Trans. Circuits Syst. for Video Technol.*, 13(7):637–644, July, 2003.

[76] T. Kohonen. The self-organizing map. *Proc. IEEE*, 78:1464–1480, 1990.

[77] I. Koprinska and S. Carrato. Hybrid rule-based/neural approach for segmentation of MPEG compressed video. *Multimedia Tools and Appl.*, 18:187–212, 2002.

[78] M. Kovac and N. Ranganathan. JAGUAR: A fully pipeline VLSI architecture for JPEG image compression standard. *Proc. IEEE*, 83(2):247–258, 1995.

[79] R. Kresch and N. Merhav. Fast DCT domain filtering using the DCT and the DST. *IEEE Trans. Image Process.*, 8:821–833, June 1999.

[80] V. Kries. *Handbuch der Physiologic des Menschen*, volume 3. Braunschweig, Viewieg und Sohn, Germany, 1905.

[81] E.H. Land. The retinex theory of color vision. *Sci. Am.*, 3:108–129, 1977.

[82] B. Lee. A new algorithm to compute the discrete cosine transform. *IEEE Trans. Acoustics, Speech and Signal Process.*, 32(6):1243–1245, Dec. 1984.

[83] M.-S. Lee, M. Shen, A. Yoneyama, and C.-C. Jay Kuo. DCT-domain image registration techniques for compressed video. In *Proc. IEEE Int. Symp. Circuits Syst.*, volume 5, pages 4562–4565, Kobe, Japan, May 23–26, 2005.

[84] S. Lee. An efficient content-based image enhancement in the compressed domain using retinex theory. *IEEE Trans. Circuits Syst. for Video Technol.*, 17(2):199–213, Feb. 2007.

[85] S.-W. Lee, Y.-M. Kim, and S.W.Choi. Fast scene change detection using direct feature extraction from MPEG compressed videos. *IEEE Trans. Multimedia*, 2(4):240–254, Dec. 2000.

[86] Y. Liang, L.P. Chau, and Y.P. Tan. Arbitrary downsizing video transcoding using fast motion vector reestimation. *IEEE Trans. Circuits Syst. for Video Technol.*, 9(11):352–355, Nov. 2002.

[87] J.S. Lim. *Two-Dimensional Signal and Image Processing*. Prentice Hall, Englewood Cliffs, NJ, 1990.

[88] C. Loeffler, A. Ligtenberg, and G. S. Moschytz. Practical fast 1-D DCT algorithms with 11 multiplications. In *Proc. IEEE Int. Conf. Acoust. Speech, and Signal Process.*, volume 2, pages 988–991, Glasgow, Scotland, May 22–25, 1989.

[89] C.-S. Lu. Block DCT-based robust watermarking using side information extracted by mean filtering. In *Proc. IEEE Int. Conf. Pattern Recognition*, volume 2, pages 1001–1004, Quebec, Canada, Aug. 11–15, 2002.

[90] L. Lucchese, S.K. Mitra, and J. Mukherjee. A new algorithm based on saturation and desaturation in the xy chromaticity diagram for enhancement and re-rendition of color images. In *Proc. IEEE Int. Conf. Image Process.*, pages 1077–1080, Thessaloniki, Greece, Oct. 7–10, 2001.

[91] P.C. Mahalanobis. On the generalised distance in statistics. *Proc. Nat. Inst. Sci. India*, 12:49–55, 1936.

[92] S. Mallat. A theory for multiresultion signal decomposition: the wavelet representation. *IEEE Trans Pattern Anal. Machine Intelligence*, 11(7):674–693, 1989.

[93] S. Mallat. *A Wavelet Tour of Signal Processing*. Academic Press, Indian reprint, 2nd edition, 2006.

[94] S. A. Martucci. Image resizing in the discrete cosine transform domain. In *Proc. IEEE Int. Conf. Image processing*, volume 2, pages 224–227, Washington, DC, Oct. 23–26, 1995.

[95] S.A. Martucci. Symmetric convolution and the discrete sine and cosine transforms. *IEEE Trans. Signal Process.*, 42(5):1038–1051, May, 1994.

[96] P. Marziliano, F. Dufaux, S. Winkler, and T. Ebrahimi. Perceptual blur and ringing metrics: application to JPEG2000. *Signal Process.: Image Commun.*, 19:163–172, 2004.

[97] J.C. Maxwell. Theory of the perception of colors. *Trans. R. Scottish Soc. Arts.*, 4:394–400, 1856.

[98] J.C. Maxwell. The diagrams of colors. *Trans. R. Soc. Edinburgh*, 21:275–298, 1857.

[99] N. Merhav and V. Bhaskaran. Fast algorithms for DCT-domain image down-sampling and for inverse motion compensation. *IEEE Trans. Circuits Syst. for Video Technol.*, 7(6):468–476, 1997.

[100] S. K. Mitra. *Digital Signal Processing*. Mcgraw Hill, New York, NY, 3rd revised edition, 2005.

[101] S.K. Mitra and T.H. Yu. Transform amplitude sharpening: A new method of image enhancement. *Comput. Vis. Graphics Image Process.*, 40:205–218, 1987.

[102] S. Moiron, S. Faria, A. Navarro, V. Silva, and P. Assunc. Video transcoding from H.264/AVC to MPEG-2 with reduced computational complexity. *Signal Process.: Image Commun.*, 24:637–650, 2009.

[103] J. Mukherjee and S.K. Mitra. Arbitrary resizing of images in the DCT space. *IEE Proc. Vision, Image and Signal Process.*, 152(2):155–164, 2005.

[104] J. Mukherjee and S.K. Mitra. Image filtering in the compressed domain. In *Proc. 5th Indian Conf. Comput. Vis. Graphics Image Process.*, pages 194–205, Madurai, India, Dec. 14–16, 2006. LNCS-4338.

[105] J. Mukherjee and S.K. Mitra. Enhancement of color images by scaling the DCT coefficients. *IEEE Trans. Image Process.*, 17(10):1783–1794, 2008.

[106] J. Mukherjee and S.K. Mitra. Image resizing in the compressed domain using subband DCT. *IEEE Trans. Circuits Syst. for Video Technol.*, 12(7):620–627, July, 2002.

[107] J. Mukhopadhyay. Isolating neighbor's contribution towards image filtering in the block DCT space. In *Proc. IEEE Int. Conf. Image Process.*, pages 2765–2768, Hong Kong, Sept. 25–30, 2010.

[108] J. Mukhopadhyay and S.K. Mitra. Resizing of images in the DCT space by arbitrary factors. In *Proc. IEEE Int. Conf. Image Process.*, pages 2801–2804, Singapore, Oct. 24–27, 2004.

[109] J. Mukhopadhyay and S.K. Mitra. Color constancy in the compressed domain. In *Proc. IEEE Int. Conf. Image Process.*, pages 705–708, Cairo, Egypt, Nov. 7–11, 2009.

[110] M.J. Narasimha. Linear convolution using skew-cyclic convolutions. *IEEE Signal Process. Lett.*, 14(3):173–176, Mar. 2007.

[111] A. Neri, G. Russo, and P. Talone. Inter-block filtering and downsampling in DCT domain. *Signal Process.: Image Commun.*, 6(Aug.):303–317, 1994.

[112] V. Patil, R. Kumar, and J. Mukherjee. A fast arbitrary factor video resizing algorithm. *IEEE Trans. Circuits Syst. for Video Technol.*, 16(9):1164–1171, Sept. 2006.

[113] S.C. Pei and C.M. Cheng. Dependent scalar quantization of color images. *IEEE Trans. On Circuits Syst. for Video Technol.*, 5(2):124–139, Apr. 1995.

[114] S.C. Pei and Y.C. Zeng. Virtual restoration of ancient chinese paintings using color contrast enhancement and lacuna texture synthesis. *IEEE Trans. Image Process.*, 13(3):416–429, 2004.

[115] T. Pevny and J. Fridrich. Determining the stego algorithm for JPEG images. *IEE Proc. Inf. Secur.*, 153(3):77–86, Sept. 2006.

[116] S. Porwal and J. Mukherjee. A fast DCT domain based video downscaling system. In *Proc. IEEE Int. Conf. Acoustics, Speech, Signal Process.*, pages 885–888, Toulouse, France, May 15–19, 2006.

[117] S. Porwal and J. Mukherjee. An integrated approach for downscaling MPEG video. In *Proc. 5th Indian Conf. Comput. Vis. Graphics Image Process.*, volume LNCS-4338, pages 686–695, Madurai, India, Dec. 14–16, 2006.

[118] N. Provos. Defending against statistical steganalysis. In *Proc. 10th USENIX Security Symp.*, volume 10, pages 323–335, Washington, DC, Aug. 13–17, 2001.

[119] M. Rabbani and R. Joshi. An overview of the JPEG 2000 still image compression standard. *Signal Process.: Image Commun.*, 17(1):3–48, Jan. 2002.

[120] R. Reeves, K. Kubik, and W. Osberger. Texture characterization of compressed aerial images using DCT coefficients. In *Proc. SPIE: Storage and Retrieval for Image and Video Databases*, volume 3022, pages 398–407, Newport Beach, CA, Feb. 1997.

[121] G. Reyes, A. R. Reibman, S. F. Chuag, and J. C. I. Chuang. Error resilient transcoding for video over wireless channels. *IEEE J. Sel. Areas Commun.*, 18(6):1063–1074, Jun. 2000.

[122] R. Rosenbaum and D. Taubman. Merging images in JPEG2000-domain. In *Proc. Visualization, Imaging and Image Process.*, volume 1, pages 249–252, Benalmadena, Spain, Sept. 8–10, 2003.

[123] S.-F.Chang and D.G. Messerschmitt. Manipulation and composition of MC-DCT compressed video. *IEEE J. Sel. Areas Commun.*, 13(1):1–11, 1995.

[124] C.L. Salazar and T.D. Tran. On resizing images in the DCT domain. In *Proc. IEEE Int. Conf. Image Process.*, volume 4, pages 2797–2800, Singapore, Oct. 24–27, 2004.

[125] C.L. Salazar and T.D. Tran. A complexity scalable universal dct domain image resizing algorithm. *IEEE Trans. Circuits Syst. for Video Technol.*, 17(4):495–499, 2007.

[126] S.J. Sangwine and Eds. R.E.N. Horne. *The Colour Image Processing Handbook*. Chapman and Hall, London, 1998.

[127] T. Shanableh and M. Ghanbari. Heterogeneous video transcoding to lower spatial-temporal resolutions and different encoding formats. *IEEE Trans. Multimedia*, 2(2):101–110, Jun. 2000.

[128] J. Shapiro. Embedded image coding using zero trees of wavelet coefficients. *IEEE Trans. Signal Process.*, 41(12):3445–3462, Dec. 1993.

[129] B. Shen, I.K. Sethi, and V. Bhaskaran. Adaptive motion vector resampling for compressed video downscaling. *IEEE Trans. Circuits Syst. for Video Technol.*, 9(6):929–936, Sept. 1999.

[130] K. Shen and E. Delp. A fast algorithm for video parsing using MPEG compressed sequences. In *Proc. IEEE Int. Conf. Image Process.*, volume 2, pages 252–255, Washington, DC, Oct. 23–26, 1995.

[131] Y. Q. Shi and H. Sun. *Image and Video Compression for Multimedia Engineering: Fundamentals, Algorithms and Standards*. CRC Press, Taylor and Francis Group, Boca Raton, FL, 2008.

[132] G.S. Shin and M.G. Kang. Transformed domain enhanced resizing for a discrete-cosine-transform-based code. *Opt. Engg.*, pages 3204–3214, Nov. 2003.

[133] M. Shneier and M.A. Mottaleb. Exploiting the JPEG compression scheme for image retrieval. *IEEE Trans. Pattern Analysis Machine Intelligence*, 18(8):849–853, Aug. 1996.

[134] H. Shu and L.-P. Chau. An efficient arbitrary downsizing algorithm for video transcoding. *IEEE Trans. Circuits Syst. for Video Technol.*, 14(6):887–891, June, 2004.

[135] T. Sikora. MPEG : Digital video coding standard. *IEEE Signal Process. Mag.*, 14(9):82–100, Sept. 1997.

[136] B.C. Smith and L. Rowe. Algorithms for manipulating compressed images. *IEEE Comput. Graph. Applicat. Mag.*, 13(Sept.):34–42, 1993.

[137] J.R. Smith and S.F. Chang. Transform features for texture classification and discrimination in large image databases. In *Proc. IEEE Int. Conf. Image Process.*, pages 407–411, Austin, TX, Nov. 13–16, 1994.

[138] K. Solanki, K. Sullivan, U. Madho, B.S. Manjunath, and S. Chandrasekaran. Statistical restoration for robust and secure steganography. In *Proc. IEEE Int. Conf. Image Process.*, volume 2, pages 1118–1121, Sept. 11–14, 2005.

[139] S. Susstrunk and S. Winkler. Color image quality on the internet. In *Proc. IS&T / SPIE Electronic Imaging 2004: Internet Imaging V*, volume 5304, pages 118–131, San Jose, CA, Jan. 19, 2004.

[140] R. Swann and N. Kingsbury. Transcoding of MPEG-II for enhanced resilience to transmission errors. In *Proc. IEEE Int. Conf. Image Process.*, volume 2, pages 813–816, 1996.

[141] J. Tang, E. Peli, and S. Acton. Image enhancement using a contrast measure in the compressed domain. *IEEE Signal Processing Lett.*, 10(10):289–292, Oct. 2003.

[142] D. Taubman. High performance scalable image compression with EBCOT. *IEEE Trans. Image Process.*, 9(7):1158–1170, 2000.

[143] P.E. Trahanias and A.N. Venetsanopoulos. Color image enhancement through 3-D histogram equalization. In *Proc. 11th IAPR Int. Conf. Pattern Recognition*, volume III, pages 545–548, The Hague, Netherlands, Aug. 20–Sept. 3, 1992.

[144] A Vetro, C. Christopoulos, and H. Sun. Video transcoding architectures and techniques: an overview. *IEEE Signal Process. Mag.*, 20(3):18–29, Mar. 2003.

[145] A. Vetro, J. Xin, and H. Sun. Error resilience video transcoding for wireless communications. *IEEE Wireless Commun.*, 12(4):14–21, Aug. 2005.

[146] M. Vetterli. Fast 2-D discrete cosine transform. In *Proc. IEEE Int. Conf. Acoustics, Speech and Signal Process.*, volume 10, pages 1538–1541, Tampa, Finland, Mar. 26–29, 1985.

[147] K. Viswanath. *Image Transcoding in Transform Domain.* PhD thesis, Dept. of Computer Science and Engineering, Indian Institute of Technology, Kharagpur, Feb. 2009.

[148] K. Viswanath, J. Mukherjee, P. K. Biswas, and R. N. Pal. Wavelet to DCT transcoding in transform domain. *Signal Image and Video Process., Springer*, (4):129–144, 2010.

[149] G. K. Wallace. The JPEG still picture compression standard. *Commun. ACM*, 34(4):30–44, 1991.

[150] H. Wang and S.-F Chang. A highly efficient system for automatic face region detection in MPEG video. *IEEE Trans. on Circuits Syst. for Video Technol.*, 7(4):615–628, Aug. 1997.

[151] H. Wang, L.P. Kondi, A. Luthara, and S. Ci. *4G Wireless Video Communications.* Wiley, Chichester, UK, 2000.

[152] Z. Wang, A.C. Bovik, H.R. Sheikh, and E.P. Simoncelli. Image quality assessment: from error visibility to structural similarity. *IEEE Trans. Image Process.*, 13(4):600–612, Mar. 2004.

[153] Z. Wang, H.R. Sheikh, and A.C. Bovik. No-reference perceptual quality assessment of JPEG compressed images. In *Proc. IEEE Int. Conf. Image Process.*, volume I, pages 477–480, Rochester, NY, Sept. 22–25, 2002.

[154] A. Westfeld and A. Pfitzmann. Attacks on steganographic systems. In *Proc. Third Int. Workshop on Information Hiding*, volume LNCS-1768, pages 61–76, Dresden, Germany, Sept. 29–Oct. 1, 1999. Springer-Verlag.

[155] T. Wiegand, G. J. Sullivan, G.Bjntegaard, and A. Luthra. Overview of the H.264/AVC video coding standard. *IEEE Trans. On Circuits Syst. for Video Technol.*, 13(7):560–576, July, 2003.

[156] S. Wolf, R. Ginosar, and Y. Zeevi. Spatio-chromatic image enhancement based on a model of humal visual information system. *J. Visual Commun. and Image Representation*, 9(1):25–37, Mar. 1998.

[157] H.R. Wu and Z. Man. Comments on "fast algorithms and implementation of 2-D discrete cosine transform". *IEEE Trans. Circuits Syst. for Video Technol.*, 8(2):128–129, Apr. 1998.

[158] H.R. Wu and F.J. Paoloni. A two-dimensional fast cosine transform algorithm based on Hou's approach. *IEEE Trans. Signal Process.*, 39(2):544–546, Feb. 1991.

[159] G. Wyszecki and W.S. Stiles. *Color Science: Concepts and Methods, Quantitative Data and Formulae.* New York: Wiley, 2nd edition, 1982.

[160] J. Xin, M.T. Sun, B.S. Choi, and K.W. Chun. An HDTV-to-SDTV spatial transcoder. *IEEE Signal Process. Lett.*, 12(11):998–1008, Nov. 2002.

[161] J. Xin, A. Vetro, H. Sun, and Y. Su. Efficient MPEG-2 to H.264/AVC transcoding of Intra-coded video. *EURASIP J. Adv. in Signal Process.*, doi:10.1155/2007/75310(ID 75310):12 (no.), 2007.

[162] C.H. Yeh and C.J. Kuo. Polynomial motion vector resampling algorithm. In *Proc. IEEE Int. Conf. Multimedia and Expo*, pages 1144–1147, Tokyo, Japan, Aug. 22–25, 2001.

[163] B.L. Yeo and B. Liu. Rapid scene analysis on compressed video. *IEEE Trans. Circuits Syst. for Video Technol.*, 5(6):533–544, 1995.

[164] B.L. Yeo and B. Liu. Visual content highlighting via automatic extraction of embedded captions on mpeg compressed video. In *Proc. SPIE Digital Video Compression: Algorithms and Technologies*, pages 142–149, San Jose, CA, Feb. 5–10, 1995.

[165] C. Yim. An efficient method for DCT-domain separable symmetric 2-D linear filtering. *IEEE Trans. Circuits Syst. for Video Technol.*, 14(4):517–521, Apr. 2004.

[166] J. Youn, M.-T. Sun, and C.-W. Lin. Motion vector refinement for high-performance transcoding. *IEEE Trans. Multimedia*, 1(1):30–40, Mar. 1999.

[167] H.J. Zhang, C.Y. Low, and S.W. Smoliar. Video parsing and browsing using compressed data. *Multimedia Tools Appl.*, 1:89–111, 1995.

[168] Q. Zhang, P.A. Mlsna, and J.J. Rodrigues. A recursive technique for 3-D histogram enhancement of color images. In *Proc. IEEE Southwest Symp. Image Anal. Interpretation*, pages 218–223, San Antonio, TX, Apr. 8–9, 1996.

[169] Y. Zhong, H. Zhang, and A.K. Jain. Automatic caption localization in compressed video. *IEEE Trans. Pattern Anal. Machine Intelligence*, 22(4):385–392, Apr. 2000.

[170] Z.Wang and A.C. Bovik. A universal image quality index. *IEEE Signal Process. Lett.*, 9(3):81–84, Mar. 2002.

# Index

T - #0837 - 101024 - C302 - 234/156/13 - PB - 9781138113787 - Gloss Lamination